MW00803703

CONNECTIONS

Social Media and Networking Techniques for Writers

By Edie Melson

Connections: Learn How to Make Valuable Connections through Blogging &
Social Networking
Copyright © 2013 by Edie Melson

Manufactured in the United States of America. All rights reserved. No part of
this book may be reproduced in any form or by any electronic or mechanical
means including information storage and retrieval systems without permission in writing from the publisher except by a reviewer, who may quote brief
passages in a review. Published by My Book Therapy, a division of Susan May
Warren Fiction, LLC, 20 Wild Plum Dr., Grand Marais, MN 55604.
First Edition Print:

ISBN-13: 978-0-9846969-9-4

First Edition eBook:

ISBN-13: 978-0-9846969-8-7

Visit our Website at www.mybooktherapy.com for information on more
resources for writers.

To receive instruction on writing or further help with writing projects via My
Book Therapy's fiction editing services, contact info@mybooktherapy.com.

Designed by: David C. Warren

Edited by: Susan May Warren

DEDICATION

*To my husband, Kirk, for his love
and for pushing me out of the nest
so I could find my wings and fly.*

Acknowledgments

For my amazing boys, Jimmy, Kirk, and John. They've always supported and encouraged me to take time to follow my own dreams.

For Alton Gansky, who first noticed I had the ability to build something through Social Media, and brought me on board to work with him.

For Susie May Warren who took a chance on me and let me become a part of this wonderful writing family, My Book Therapy. I wouldn't be where I am without this team and their friendship. Thank you Beth, Reba, Lisa, Rachel, Alena, and Michelle.

For my critique partners, who've stuck with me throughout the years. Thank you Vonda, Mary, Pam, and Caroline.

And most especially my Lord, who gave me this crazy wonderful gift of writing.

"For I am confident of this very thing, that He who began a good work in you will perfect it until the day of Christ Jesus."
Philippians 1:6 NASB (New American Standard Bible)

CONTENTS

Part One — Blogging: Developing Content

Part Two — Blogging: Promoting your Content

Part Three — Social Networks: Facebook, Twitter and Developing Relationships

Part Four — Social Networks: Afterthought Platforms that shouldn't be Afterthoughts

Part Five — Knowledge & Tools Bringing it all Together

INTRODUCTION

"So, how much time do I really have to spend on social networking?"

Anytime I'm introduced as an expert on social media I can count on this question coming up.

"Truthfully," I always answer in a low voice so they have to pay attention to catch what I say, "the bare minimum."

My answer never fails to bring shock and amazement. People assume that since I'm successful with social media, I'm going to insist they dedicate their lives to it. Or they think they'll have to spend hours to get the same results as me.

Both assumptions are dead wrong.

But what about the hype, the promises, the RESULTS? Don't the results increase in direct proportion to the effort? No, not so much.

Fairly quickly, the return on investment when it comes to time and social networking, begins to diminish. I know that sounds counterintuitive, but it's the absolute truth.

That said, how much time is required? It depends...on your expertise, your audience and the desired result.

Your Expertise

With any new skill, including social networking, there is a learning curve. It takes time to come up to speed on how to use Twitter, Facebook, and blogging platforms effectively. But, you don't need to become an expert. After all, you want to be a writer—not a social media expert! Here are the skills you need to see results with social networking. (I like to call these the Marketing Triad.)

Blogging

You should have a blog—to practice writing on a deadline if nothing else.

You need to know how often to post.

- How to use keywords and labels effectively.
- How to tie your title to your keywords.
- How to use images and videos to illustrate and partner with your posts.
- How to answer comments effectively.
- Which platform best suits your needs and skill level.

Facebook

You need to have a Facebook account and know how to navigate Facebook.

- How to accept friend requests, as well as send them.
- How to hide unwanted posts.
- How to compose an effective Facebook post.
- How to configure your account privacy settings to protect yourself and your family while still interacting with readers and clients.
- The difference between a regular page and a fan page.

Twitter

You need to have a Twitter account and know the basics of TWEETing.

- How to compose an effective TWEET.
- How to use HASHTAGS correctly.
- The difference between a direct message, a reply, and a TWEET.

Conclusion

For now here's my short answer on how much time to spend each day once you're familiar with social networking. No more than 30 to 45 minutes per day, five days a week. After talking with thousands of authors and writers, I've found that any more time than that becomes counter-productive and actually interferes with our goals. After all, we're at the computer to write, not socialize! This book is written in a conversation style, as if we were having coffee to help you tackle these topics one by one, in depth. At the end of each conversation, you'll find a "Connect!" activity to help you implement these concepts and ideas and move you on the road to creating connections.

Blessings,
Edie

Part One — Blogging:

Developing Content

WRITING FOR THE INTERNET AUDIENCE

There's a mountain of misconceptions about writing for the web. Let's spend some time debunking and discussing them.

Even today, people assume that everyone reads the same way, no matter what the medium or delivery system. That kind of assumption can kill a blog quicker than anything.

One of the great things about the Internet is the ease with which people can find information. Readers are looking for information, and they expect to find it quickly. They want to be able to read or scan the content to find what they are looking for. The Internet audience isn't willing to wade through pages of, to them, worthless information to get at what they need. That means the author must make organization and readability of primary importance.

The writer must also take into consideration the delivery system of their medium. Studies show that people generally read up to twenty-five percent slower on computer screens. The reasons are complex, but here are a few:

- Computer monitors are harder on the eyes than paper.
- They generally have fairly low resolution, so the words aren't as sharp.
- Also, while the contrast between ink and paper is usually strong and fairly consistent, monitor settings can vary widely depending on type and settings.

Because of these factors, most people find it tiring and even frustrating to read long articles online. But there is a way to capitalize on the medium.

People want information, so give it to them—up front. This is called Front-loading the paragraphs. In the past, writers were taught to work up to the information, building a case for its validity. There would be a theme sentence and opening paragraph, followed by a well-reasoned article, ending with the conclusions. That was known as the Pyramid style of writing. Instead, Internet writers state the conclusion and most important information first, followed by the reasoning that got them to the point. Writing for the Internet turns the Pyramid upside down.

Writing for the Internet also turns the standards of formatting upside down. Anyone who's done any writing knows how to format a presentation. Utilize the font Times New Roman, 12 point, single-space the document, and indent for paragraphs. Not so on the Internet.

All of those rules are incorrect when writing for the web. Not only are they incorrect, they'll keep your blog from growing and in some instances, from even being read.

Learn the correct way to format

You're writing for the reader who wants to scan. This means including bold headings, lists, and bullet points. The text needs to be easy to read, and you'll need to minimize eyestrain.

Here are the things you'll need to do:

- Wide margins. There should be approximately 12 words per line, max.

- If your text is longer than one to two printed pages, try to break it up into separate web pages. It'll be easier to read and the pages will download faster, especially if the user has a dial-up connection.

- Avoid a busy background or frame.

- Consider the contrast between text and background. Although white is an okay choice for background color, an even better one is a shade that is lightly pigmented. Think pale gray, peach, or green. These are much easier on the eyes.

- Choose your font wisely. Times New Roman isn't a good choice for reading on the computer. Arial, Helvetica, and Verdana are better choices. Also take into account font size.

- Don't indent paragraphs; instead skip a line between them. This gives the reader's eyes a chance to rest. This is called block formatting.

- Keep the paragraphs short, no more than 100 words.

The final part of the equation you need when formatting for the Internet audience is the graphic elements. This includes much more than just images, but also videos.

In years past, the emphasis with writing was simply that—writing. Now, as our society has become more and more visual we, as writers, must also evolve.

This is especially true on the Internet. We must broaden our horizons and become designers. Trends and statistics are clear; in less than five years, eighty-five percent of what is viewed on the Internet will be video driven. This means we need to illustrate our blogs with images. As writers, we often view our words within a box, at least our minds. We don't pay attention to the whole picture. We can no longer afford that mindset. When writing for the web, we have to educate ourselves.

- Look at the web pages you go back to and make a list of what catches your eye.

- Look at web pages that frustrate you and make a list of your frustrations.

- Notice what entices you when you're reading a magazine article.

All of these tools will help you become a more savvy blogger. Beyond that, make a commitment to stay current on the trends in your business—the Internet and all things digital.

Choosing Your Weapon

That question—or some variation—seems to come up frequently when I speak about blogging. And the answer is simple—and complex.

First and foremost, I strongly urge you to stick with one of the top three blogging platforms—Blogger, WordPress, or Typepad. All of these are good, solid platforms with the plugins you need to build a long-term, healthy blog. Personally, I prefer Blogger. The reason I started with Blogger was because it was so easy. And when I started my blog, I could hardly write a complete sentence about what I knew of blogging, much less a book!

Blogger

I'll be honest. I have an affinity for Blogger because we've sort of grown up together. It seems that every time I'm ready to take that next step with my site, Blogger has added a new upgrade.

Blogger is run by Google. Now Google might not be perfect, but they have a HUGE presence on the Internet and they—as well as anyone—know where this business is headed. I am familiar with WordPress and also with Typepad as blogging platforms. And I don't want to leave them out. They both have some great options as well.

Connections

WordPress

The thing I like best, and least, about WordPress is its tutorials. It walks you through the mechanics of building a blog. But it's a lot to get through to just get started. When I built my blogs, it was harder to get started with WordPress if you didn't know anything about websites and such. Now with their directions, it's pretty easy. Their tutorials are much more thorough than Blogger's. Blogger walks you through the steps; WordPress explains them.

WordPress is also constantly updating its options. As a matter of fact, my next blog will be built on WordPress. Not because I'm unhappy with Blogger, but because I want to expand my knowledge base and play with a new toy!

Typepad

This platform isn't free and it's my least favorite, maybe because I'm not as familiar with it. But I'm coming up to speed quickly because I've taken over joint management of a large Typepad site.

Surprisingly, since it's a fee-based service, it seems to have the least options. Every time I want to do something, it's only possible in the next pricing tier. Everyone has their favorite platform and to be truthful, they're all pretty good. WordPress and Blogger are definitely the frontrunners, and they tend to swap the lead back and forth. For a while, Blogger offers the newest and best options, then for a while WordPress does. I don't think you can lose with either one.

Connect! Activity

Spend some time visiting blogs, looking for things you like and don't like. Study the sites you return to time and again. Notice what platforms they're built on.

Here are some things to consider:

- How easy is it to navigate the blog?
- How easy is it to connect?
- Does the layout make sense?
- Do the color choices make it easier to read, or harder?
- How often do your favorite blogs offer new posts?

Is Blogging the Right Choice for Me?

For several years, blogging has been touted as the best way to reach an audience—especially for writers. Now, enough time has passed that it's possible to get a good statistical foundation of whether or not blogging really is a good use of a writer's time.

The answer is...it depends.

I know. I can hear the groans from here, and I feel your pain. Everyone was promised, "Blog and they will come." Well, that is only partly true, and here's the nitty-gritty of blogging.

I do believe writers need to be affiliated with some type of blogging, but a personal blog may not fit your style. If that's the case, consider a group blog, or a blog where you have a regular column. It's important, in today's market, to have somewhere you can engage your audience in a conversation.

Blogging works for writers under these circumstances:

•   The blog/blogger has a clear purpose to blogging.

•   The audience is clearly defined. (For example, a novelist is writing for his readers—not other writers.)

•   The blog demonstrates who the writer is. Or, in other words, it enhances—not contradicts—the author's brand.

•   The rest of your branding makes sense with your blog.

Connections

Blogging DOES NOT work for writers under these circumstances:

- The point of the blog is vague and undefined.
- The audience isn't clearly defined.

The blog leads readers to a different understanding of the writer—not a deeper picture—but totally different. That said, should you have a blog? In this section on blogging, I'm going to give you the tools to answer that question for yourself. Because in reality, this isn't a one-size-fits-all world.

Connect! Activity

Think about the things that interest you, things you're passionate about. Then do some online digging, mining for your own personal blogging gold.

Here's what to look for:

- Research what's already out there in regard to your passions.
- Investigate online groups that are natural gathering places for those who are interested in the same things.
- Look at the most popular sites in your areas of interest. One thing you don't want to do is reinvent the wheel. Instead, build on what's already available.

FIND A FOCUS FOR YOUR BLOG

For years, there has been an unwritten—oft repeated—Rule of Speaking. Tell the audience what you're going to say.

Say it.

Tell the audience what you just said.

Even though this formula is a cliché of epic proportions, there is a foundational truth of communication found within it. It's vital that your audience knows what to expect before they'll invest time reading what you have to share. People are busy and they have to make choices about where they'll spend their time. It's your first job to give them the information they need to make that choice.

So how do you find a purpose for your blog?

- Ask yourself what you want to accomplish from it.
- If you want to share your passion, start with that.
- Do you want to make a difference in the world? Ask yourself what you think would make the world a better place.
- Is your purpose just to build a platform? Then do some research and decide who will be interested in your platform and go after those viewers.

Focus is of paramount importance when it comes to blogging. An unfocused blog is an unvisited blog.

Once you have your primary purpose, it's time to refine it. Think out of the box and find a niche or a unique perspective that hasn't been explored yet.

Connections

I'll use my blog as an example. I knew I wanted to provide a blog to teach and empower writers. Most writing sites also incorporate book reviews. How could I tweak that to be original and fresh? The answer I found was to review books and tools for writers. Keep refining, even after your blog is live. That's one of the advantages of blogging—you can respond to changes in what your readers want and is changing the market.

Defining Your Audience

This is vitally important to your blog's viability. A poorly targeted audience is a frequent reason for a blog's lack of success. The target of your blog may seem like an obvious thing, but I'm continually amazed at the number of writers who get off track here.

Common Misconception: I'm a writer, so my blog should target other writers.

Truth: Unless you're teaching writers about craft, your audience isn't other writers.

The Fix: This goes back to the purpose of your blog.

Many novelists use a blog to connect with their readers. If that is the purpose of your blog, then your audience is your reader—not other novelists.

So what does this look like in application? For a suspense writer, it would look like Brandilyn Collins's site (www.brandilyncollins.com). She does a great job of connecting to her readers. She also has a separate site that targets other writers because she is a popular conference instructor and teaches others the craft.

Suppose you're a freelance writer? Then your site should showcase your work and connect you with the people you write for. My site, for instance, targets businesses that need someone to write web content. This provides me with the opportunity to teach writers the craft of writing for the Internet and proves my expertise to potential customers.

Your Blog and Brand

Your blog should make sense with your brand. It will help complete the picture of who you are. This goes hand in hand with defining your audience. It also takes into account the purpose of your blog. All of these things must work together to present a clear picture of who you are and what you have to offer your audience.

Trying to figure out your brand can sometimes be difficult. Here are some things to consider when choosing your brand:

• What is your passion? You'll only be able to sustain a blog about something you're passionate about.

• What do you know that could help someone else? We all have things we're good at.

• What do you want to become an expert about? If you want to learn about something the best way to do it is to teach someone else.

The final thing to do is get started. All blogs evolve and get better with time. All the research in the world can't take the place of actually doing the work. So don't hesitate, take a chance and try blogging!

Connect! Activity

In the previous Connect! activity, you spent time online doing research. Now it's time to put that research into practice.

• Brainstorm ways you could approach your passion from another direction.

• Consider ways to connect with others who share similar interests.

• As you browsed those sites, what gaps did you notice? For instance, when I wanted to add book reviews to my site, I noticed there weren't any sites strictly reviewing writing books.

• Begin to write blog posts. Don't post them. Instead, store them in your word processing program. Write a minimum of four posts per week for four weeks.

• After that four weeks, go back over what you've written. Look over the themes and topics. You will find the passion you're seeking. It's the topic you keep coming back to and keep exploring in new ways.

How to Find Ideas for Blog Posts

It happens to us all. It's the night before a blog post is due, and it seems like all our creative juices have run dry. Do we just bail and skip that post, or is there something constructive that can help us out?

Don't panic. There are some specific things you can do and places you can look to find, not just something to fill in with, but something truly creative to blog about.

First, look no farther than your last blog post, or more specifically, the comments your last post elicited. Your subscribers and readers are your best source of inspiration. Their questions and comments can spur you on to ideas you'd never have considered otherwise.

Second, take a look at the blogs you follow. What are some of the questions you've asked your favorite bloggers? What are some things you wish you knew? Just because you don't know the answer, doesn't mean you can't write about it. Trust me, the best way to learn something is by researching it thoroughly.

Third, investigate Facebook. Do you see any trends, or notice a subject everyone is talking about? There's a blog post in there!

Fourth, have you recently solved a problem that has to do with your blog's topic? Don't be stingy. Share your secret with the rest of us. If you had an issue or challenge with something, chances are you aren't the only one. You could end up being the blogging hero of the week.

Connections

Fifth, notice what day it is. Not just the day of the week, but the month. For example, did you know June 26 is International Day Against Drug Abuse and Illicit Trafficking? Also, in the same vein, what season is it? Every season opens up all sorts of ideas for posts.

Sixth, at a loss for words? Post a picture or a video or even a cartoon instead. (Just make sure you're not violating any copyright laws.)

Seventh, post a question or a rant about something that truly irritates you.

The point of your blog is to open up the lines of communication—to get the conversation going. Digging your posting topics out of things you're interested in can grow your blog in ways you never imagined.

Connect! Activity

This activity will help you begin an idea file system for your blog. If you don't start now, you'll lose some of those great ideas or find yourself at a loss when it comes time to post. Start with a file folder. It can be on your computer, or in your office. Just make sure you have EASY access to it.

It should contain:

• General blog ideas or topics.

• Calendar ideas. During the year, you'll discover all sorts of interesting days. Frequently it will be on that day, too late to write a post. Make a list and you'll be ready for next year.

• Innovations that have to do with your blog's focus.

• Sites that have good ideas. You don't want to steal their posts, but you could write about similar things for your specific audience.

• Pictures to illustrate your blog. These need to be digital pictures that you can use without copyright infringement. I recommend you keep this file on your computer. Don't assume because you can access a photo online today that it will always be available. Sites change and so does their content.

• Within the picture file, also keep a list of photos you wish you had. If you're a photographer, look for those specific shots. If you're not, ask a friend who is.

How Often Should I Blog?

There is lots of information about blogging around, some of it good and some of it not so good. One of the biggest misconceptions I see is in regard to how often you should post your blog. The most common myth seems to be that daily is always best. While this may be true for some, it's definitely NOT true for the majority and here's why.

• Daily blog posts don't always get read. People are busy and few (if any) have time to read every blog every day. If you post once or twice a week readers are more likely to read every one than if you post daily. For the sites I follow that post daily, I pick and choose what I read by the title of the post (which is another subject).

• A daily schedule can also cause burnout. For many people, a blog is something they do in addition to their regular writing. If you have to work too hard at your blog, you are more likely to quit.

• The posts are frequently better when they're spaced further apart. Let's face it; it's hard to be great day-in and day-out. This definitely holds true with blogging.

So what is a reasonable blogging schedule? Truthfully, it depends. But here are some things to think about to help you find that magic number for you.

• **Your personality—** Are you easily discouraged when you don't live up to expectations? If so, I'd start slowly. Once a week is a good goal and then, if that's manageable you can add another day. But keep the schedule in line with what you can accomplish. If you're like me, when I fail at something I often give up and quit.

Connections

- **Your goals—** Think about what you want your blog to accomplish. Are you trying to connect with your readers, or start a network? These will entail posting more often and replying to comments in a timely fashion. Do you want to build a platform? With that, a regular schedule of two to three times a week, with focused articles and posts should do the trick.

- **Your lifestyle—** what does your home life look like? Do you have a full-time job or are you a full-time writer? I have a friend who writes, and she homeschools five kids. Believe me, posting four or five times a week just isn't a reasonable goal for her.

So what's the answer? The key is to be flexible. Give yourself the time and space you need to figure out what works best for you and for your readers. When you do that, you'll be well on the road to excellence.

Connect! Activity

As I mentioned, you'll want to take some time developing a workable posting schedule. Here are some things to consider:

- Take a realistic look at the time you have to spend on the computer. You want to make certain the majority of that is spent writing. Only spend a small fraction on your blog...unless your blog is going to be your primary source of income.

- Looking at your schedule, find the easiest days of the week to do work. For example, are you only going to be able to write blog posts on the weekend? Then you'll want your primary blog post to be on Monday or Tuesday.

- Describe or name your primary post of the week. Keep it to no more than a sentence. Shorter is always better.

- Make a wish list or idea list of other post topics you could add when and if you have the time.

TIPS ON EDITING YOUR BLOG POSTS

Lately, I've noticed more typos creeping into my blog posts. It's not a fun thing to admit, but it's because I've become complacent about editing. Oh I could blame it on being busy, or life getting crazy, but the truth is... I've gotten sloppy.

It's easy to do. Things are going well with my blog, and my concentration has been elsewhere. Still, no excuse.

Here are some tips on an efficient way to edit your blog posts. It's not a long involved process, but it will ensure those annoying typos become much less commonplace.

First, I compose my blog posts in a word document with NO formatting. **I do this for a couple of reasons:**

- It gives me a backup of everything I post on my blogs.
- It's easier to check for misspelled words here, rather than in the compose pane of my blog.

Connections

Blog Edit Check List

I begin the process by looking for misspelled words. First checking for the ones my word program underlines in red, then checking some common words that have more than one spelling. **Here's a short list of ones that frequently slip by me:**

- lose vs. loose.

- chose vs. choose.

- its vs. it's.

- there vs. they're vs. their.

- your vs. you're.

At this point I stop to copy and paste the post into the compose window of my blog. This is when I add the formatting. The reason I wait is because the formatting doesn't copy and paste accurately because of the code involved with publishing a blog to the web. I check for places to break up the text with bold face, headings, italics, and bullet points.

Next, I add the photographs to illustrate the post.

Now that I have the photos, I go to the preview window to see if the font type is correct and all the spacing is accurate. **Here's what I'm looking for:**

- Block formatting—Single-spaced paragraphs with an extra line between paragraphs.

- An extra line between bullet points.

- No other extra lines anywhere.

- That the text and the pictures line up well and there isn't an odd or short line of text sticking out anywhere around the photos.

Finally, I return to the compose window, and read the entire post out loud. I know it sounds weird, but your brain uses different pathways when you read something aloud, and you're more likely to see what's actually on the page rather than what you intended to put there.

These steps won't ensure a perfect post every time, but they will cut way back on accidental typos.

Connect! Activity

It's important to have a quick method of checking your posts before you hit the publish button, especially when you're first starting out.

Here's how you can develop one specifically for you:

•   Make a list of words you tend to overuse when you write. For me, "actually" is at the top of the list. I think I could write an entire post and use that word in every single sentence. Not a good thing!

•   Make a list of your own common mistakes or things you need to correct. I never seem to write with contractions, I always spell out the individual words. So one of the first things I do is go back through a post and look for places to add contractions.

•   Make a list of the buzzwords or slang that have to do with your blog's focus.

COMMON BLOGGING MISTAKES TO AVOID

Here is my list of mistakes I see bloggers making on a regular basis:

• **Too Little Visual Contrast**— Many times bloggers forget about contrast when they choose colors for their blog. Make sure that your content (text) is easily readable against the background color you've chosen.

• **Too Busy**— Blogs are often so crowded with plugins and other stuff, like add-ons to a post. When you end a blog post, end it. Don't tack on extra announcements or anything else. If you do, readers may get distracted and be unable to figure out where the body of your blog is. Remember the focus of your blog is the post. If the reader can't find it, something's wrong.

• **Too Long**— Effective blogs are short. I know, we're writers and we love words. But, reading on a computer is hard. Be considerate of your reader, and keep your blog post short and focused.

• **Too Much Stuff after the Post**— If you include other stuff after your blog post, stop it! I've been on blogs where a post ends, but before you can get to the comments option you have to scroll through advertising, announcements and other extra stuff. If you need that stuff, add another post to your schedule. Don't just fill up the space with everything once a week because you're trying to save time. Your reader is busy too and doesn't want to wade through unrelated things just to comment.

• **Wordy Sentences**— While we're on the subject of length, also watch out for sentence length (no more than 12 words long) and paragraph length (no more than 50 words long).

Connections

- **Wrong Font/Size—** Again, reading on a computer is tough on the eyes. Don't make things worse by using a font like Times New Roman. Stick to simple fonts like Verdana or Arial.

- **Incorrect Formatting—** Blogs should be written with the reader in mind. This means using block formatting. Don't indent paragraphs, but add an extra line of white space between paragraphs.

Connect! Activity

Rather than going over your blog by yourself, this is the time to enlist the help of a friend. Sit down together and review this list to get someone else's opinion. It's often hard for us to evaluate something we've developed.

WHAT DEFINES SUCCESS?

Considering all these things, you need to know how to measure success as a blogger. Many times, beginning bloggers have an unrealistic idea of how fast their blogs should grow.

There are lots of exceptions to how fast a blog can grow. We can all find examples of blogs that have grown incredibly fast, but we should never judge the growth of our own blogs by the exceptions.

That said, I do believe there are reasonable expectations for how fast your blog audience should increase. That growth is predicated on certain variables:

• The predictability of blog posts—A blog that posts regularly will attract followers much faster than one that posts sporadically.

• The frequency of blog posts—A blog with fresh content on a daily basis will usually earn more repeat traffic than one that only offers new content once a week.

I'm a huge fan of growing your blog organically through relationships and targeted social media. Frequently, this will cause your blog to grow slower during the first year or so, but this will give you a solid base of readers and tend to speed growth in the following years.

Stage One

In this technological age, networking can usually jumpstart a blog with 20–40 followers at startup. These followers are your first foundation, but not all of them will be part of your permanent foundation. These are friends and associates who want to help a fellow entrepreneur get started. They're a great help because they'll spread the word to their friends and associates who will comprise your foundation.

This means your first six months to a year will see little forward momentum. You'll gain new followers, and lose some of the original ones. It will feel almost like two steps forward and three steps back. But this is a critical time because you're cementing the core of your audience. I think of this as gathering the snow and solidifying it into a snowball.

During this time, many bloggers get discouraged from the slow growth because they don't understand what's happening. When I talk with someone in the first year of their blog, I try to give them insight into this process so they can watch for it and rejoice as it happens.

Stage Two

After stage one comes six to nine months of small but consistent growth. Your blog has enough history at this point to have a proven track record of consistent, valuable posts. This makes your core group more willing to share your site with others.

During this stage, search out valuable guest posts. Find people you respect and invite them to write a post or ask permission to repost one of their old posts. Think of this stage as beginning to roll your snowball through the snow, gathering a more solid ball that will hold together when you roll it down the hill.

Stage Three

This is when your blog really starts to take off. Your blog's audience begins to grow a lot faster and you'll spend less time promoting yourself on social media because others will be doing it for you. They'll be talking about your blog because it's valuable to their followers and friends, not just because they like you personally.

Soon, you'll find yourself asked to guest post on other blogs, and you'll be asked for permission to repost your older blog posts. I think of this stage as when you push your snowball off the hilltop and it begins to gain momentum on its own.

An Interim Stage (can hit at any time)

During this downhill stage, you'll still hit roadblocks and times when you have to give your snowball a push. The key is to stay flexible, continue to listen to your audience and don't let up on the interactions.

Remember, this is an organic process and these stages are just loose guidelines of how the growth of a normal blog should be measured. If you're neglecting one or more of the following things in your blog plan, your blog will probably see slower growth.

- A regular posting schedule, with a minimum of one post per week.

- The consistent use of Social Media, especially Facebook and Twitter.

- Constant interaction between you and your audience by answering comments and visiting their blogs and leaving comments.

Connect! Activity

If you already have a blog, take a look at your audience. With Blogger or Typepad you can go into the stats section of your dashboard. With the free WordPress blog, you'll need to pay more attention to the comments section of your site.

Here's what you want to look for:

- Your most popular blog posts. Judge by visitors, views, comments and if it's shared on social media.

- The day of the week your blog gets the most hits.

- The types of posts that get the most comments.

FIVE REASONS YOUR BLOG ISN'T GROWING

It is easy to get discouraged with the way your blog is faring. Below are five reasons your blog may not be growing. These things are critical to the success of any blog, no matter what the focus. Take a few minutes to see how your site holds up to review.

• **Keeping a Regular Schedule.** This is one of the first stumbling blocks to blog growth. If you're not a big name in your industry, you need to be blogging on a schedule. We want our blog followers to stop by on a regular basis to visit and leave comments. They should be able to expect the same kind of regular commitment from us.

• **Answering Your Reader's Comments.** If someone is willing to take the time to leave a comment on our blogs, the least we can do is acknowledge them. Reader comments don't always have to be answered, but we can thank them for stopping by. If your blogging platform doesn't allow you to reply to individual comments, or even if it does, it's fine to group several acknowledgements together in one reply from you.

• **Social Media Interaction.** Do your readers know how to follow you on Twitter and interact with you on Facebook? How easy are these buttons to find? I can't tell you the time I spend looking for your Twitter names when I want to give you a shout out on social media. Most people won't bother searching, and you'll have missed meeting new and potential readers.

• **Do Your Blog Followers Have a Way to... Well... Follow?** Most of the blogs I visit are much better at this than in the past, but I still see a sizable number that miss this critical component. Every blog needs a Follow by E-mail

Connections

and Follow by RSS option. It doesn't matter that you don't know what RSS is, those who use it to read blogs are passionate about it and won't cross over to receive blog updates via e-mail. (See "Plugins and Gadgets" for more information about e-mail updates and RSS feeds.)

• **Are the Components Above Easy to Find?** The way you organize the sidebar of your blog is vitally important to blog growth. You should have things in the order of importance. For example, on most blogs you'll see the first options for readers are the Follow by E-mail and Follow by RSS actions. After this is my Follow me on Twitter button, etc. People are much more likely to spend money than time, so you have to make this easy for them to come back and visit regularly.

Connect! Activity

After looking at the list above, you may feel overwhelmed by the work you need to do and the changes you want to make. But don't try to do everything at once. Make a list of things you want to accomplish and then prioritize it. Tackling it one project a time will help you gain traction and stay on track.

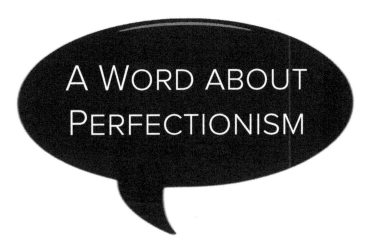

A Word about Perfectionism

I want to give you a friendly warning about something that attacks many bloggers: **the Dreaded Diva Disease!** I'm sure many of you are unfamiliar with the name of this malady, but when I describe its symptoms you'll recognize it for what it is.

A diva, in today's language, is someone who is high maintenance, who has to have things just so in order to be satisfied. Many of us fall into that trap with our blog posts. We feel our grammar, syntax, and style must be perfect before we click publish.

Actually that is the sort of stress that defeats most bloggers. I'm not advocating sloppy writing. But, I am on a mission to get some of us to relax a little. Blog posts are supposed to be short, to the point and above all, conversational. And let's face it; we all sometimes speak in sentence fragments.

Besides all that, mistakes in this medium are easily correctable. It's not like we'd have to reprint a book or magazine.

Being a perfectionist with a blog post is a colossal waste of time. I know, we all want our posts to go viral and be referred to for years as a masterpiece, but that is highly unlikely. Day to day blogging is almost disposable writing—it's here for a day or two and then gone—left to linger in the archives. Of course, savvy bloggers resurrect old posts, clean them up and use them over and over again, but that's a later subject.

That isn't to say that blogs are without benefit, but it's the sum of what we say and how we say it that proves our value to our readers.

Connections

Connect! Activity

- **Set a time limit.** Blogging isn't something we can invoice a client for, so we shouldn't spend too much time and effort in its implementation. It can augment our income by showing what we can do and proving our worth as a writer, but it isn't a direct income source for most of us.

- **Find a system to writing your blog, and use it.** Here's how I compile a blog post: First I write it out in a word document, then I read it out loud, finally I do a quick grammar and spelling check. After that, I publish it and move on to my next task. Find your system and stick to it. This will make it easier to maintain a workable schedule.

- **Post more often. Yep, you read that right.** If you're agonizing over every word and nuance in your post, you have too much time on your hands, and you need to step up the frequency of your posts. This will do two things: it will force you to spend less time because your deadlines are tighter, and it will require you to relax your standards.

Part Two — Blogging:

Promoting your Content

PLUGINS AND GADGETS EVERY BLOG NEEDS

When most people think about blogging, they concentrate on the posts themselves. As important as the posts are, just as important are the gadgets and plugins found in the sidebar.

These bits and pieces make up the roadmap that brings people back to your blog. Beyond that, often they enable you to connect on a more meaningful level with your audience.

E-mail & RSS

First are the two that allow people to subscribe to your blog. In the next section you'll learn the important difference between subscribers and followers. Subscribers actually receive notification in the e-mail inbox or RSS reader every time you put up a new post.

It's vital that you have both a way to sign up to follow by e-mail and to follow through RSS. Think of these two like the old Mac versus PC debate. You love one or the other and you're probably not willing to change. It's the same way with e-mail and RSS. People use one or the other and they are not willing to change. So have a way to sign up with both.

E-mail & RSS signup: If you have a Blogger site, there is a gadget for this. If your blog is hosted by another platform you'll want to use Feedburner for this. Feedburner is owned by Google, so this is the program that also runs the Blogger gadget.

Connections

Follow Me on Social Media

It's important for us to interact with our audience in ways beyond our blog—that means social media. **You should have a way for people to connect with you through:**

- Following you on Twitter.

- Following you on Facebook.

- Liking your Facebook Page.

- Following you on Pinterest.

- Adding you to their circles on Google Plus.

- Following/Liking you on any other major social media platforms you're a part of.

Beyond these, you also need to make sure your name is prominently displayed somewhere at the top of your blog. It can be in the title, the description, or contained within your social media area. I don't care where you put it, just so long as it's visible in the first two inches of your blog. It's awful to visit a blog and have no idea who writes it.

Now that you know you need these plugins and gadgets, there's one more thing to think about—the order you arrange them in. Here's the way I handle that: I ask myself, what is the first thing I want my readers to do? My answer is that I want them to return to my blog. So, I have my e-mail signup and my RSS signup at the top of my sidebar. Next I want them to follow me on Twitter and "like" my Facebook page. So those are the next two gadgets that go in. I use this method of deciding where to put things all the way down.

Connect! Activity

Your blog may be in good shape in regard to plugins, or you may have some work to do. In either case, a blog is always a work in progress. Here are some things to do to keep moving forward instead of bogging down.

• Look at the sites you visit most. What do you like about them? What plugins make your life easier when you're on them?

• What are your biggest pet peeves when you visit a blog? Make sure you're not guilty of the same crime.

• Talk to your friends and see what frustrates them when they visit a blog. This is a great question to ask on Facebook or Twitter. You'll get lots of good answers, and it will give you more visibility on social media.

Blog Followers vs. Subscribers

I want to clear up an issue many are confused about—the difference between a blog *Follower* and a blog *Subscriber*.

Often times the two are used interchangeably. They are not the same, and beyond that, one is extremely valuable and the other less so. It's also likely that the two numbers differ greatly. For example, I have 137 followers on Google Friend Connect and well over 400 subscribers. But before we go any further, here are the definitions:

A Follower

This is unique to Blogger sites and looks like a box with rows of pictures. If you wanted to become one of my followers, you'd click the join this site icon. Then, if you have a Google account with a picture, your face shows up on my blog. It's nice to see that people like my blog and this is a way to see that others like my blog.

Also, if you're familiar with your Google Reader, my blog shows up there. But, and this is critical, those who follow my blog this way do not receive any kind of e-mail notification when I post something new on my site.

A Subscriber

These are people who sign up to get notification of new posts on my blog through their e-mail account or through RSS. The most common way of doing this is through Feedburner.

Connections

These people are infinitely more likely to visit my blog on a regular basis. And beyond that, they're willing to have my notifications clutter up their inbox. In these days of e-mail overload, this is a big commitment.

Bottom Line

While there's nothing wrong with Google Friend Connect, make certain you also have a place for people to receive notifications through e-mail and RSS. These are your subscribers, and when a publisher is looking at your platform these will be the numbers they're interested in.

Feedburner Basics

One thing that's important to stay on top of with your blog is the stats regarding your subscribers. Notice I said subscribers, not followers. Remember, there is a distinction, and the difference is big—really big.

A follower, at least in Blogger, means they click on a follow button and their picture appears on your blog. Anyone can be a follower and never look at a blog again. A lot of folks do it and forget about it. It's a nice thing to do and a growing number of followers looks impressive—at least to those who don't know any better.

A subscriber is someone who signs up to get your blog updates in their inbox or through an RSS feed. This is a person who is willing to make a commitment. They may not always read your blog, but they do like to keep hearing about it in their inbox. These numbers may or may not show up visibly on your blog site, but these are the numbers publishers and others are interested in.

To add a Follow by E-mail or Follow through RSS, just use the plugin or gadget for this purpose. Be certain you use Feedburner. There are other options out there, but they're not as user friendly.

Adjusting your Feedburner settings is easy once you know where they're located. In Blogger, visit the webpage, or visitor side of your site (not the dashboard). Look for the e-mail subscription pane and click on the words, Feedburner. This will take you to the Feedburner settings area.

Special Note: Some blogs no longer have the Feedburner link in this place. If you don't see it on your blog, you can go directly to this link: **http://feedburner.google.com/** and sign in. Your blog should automatically pop up because Feedburner is owned by Google and everything made by Google is connected.

Once there, click on the blog (if you have more than one) you want to adjust the settings on. Across the top, you'll find the following tabs:

ANALYZE

OPTIMIZE

PUBLICIZE

MONETIZE

TROUBLESHOOT (*I guess they couldn't find anything to end in "ize" for this one*)

Under publicize you'll find the way to manage your subscriptions and discover who's subscribed to your blog. You'll also see options that will help you post your blog to other social networking sites.

I've had a lot of questions about how to schedule the exact time your blog is emailed to subscribers. For that, click E-MAIL SUBSCRIPTIONS and then DELIVERY OPTIONS.

Connect! Activity

I recommend you spend some time getting familiar with your specific subscriber management system, whether it's Feedburner or something else. Try clicking on the different options and learn what it all means. Many of these systems offer you ways to track the stats of your blog. Knowing your resources will help you increase the number of people reading your blog on a regular basis.

CONQUERING KEYWORDS AND SEO

To explain **SEO (Search Engine Optimization)** and **Keywords** effectively, I'm going to introduce you to an imaginary businessman: I'll call him Joe. He's a plumber.

Joe's a smart business owner and named his business, **ABC Plumbing**. He chose carefully because with that name, he had a good chance of being the first listing in the directory under plumbers. That was ten years ago and that ad in the telephone directory kept him supplied with customers while it made him visible around the community. But as time went by, fewer people looked at the Yellow Pages when they needed a plumber. Instead they searched online.

Joe is a with-it sort of guy, and he had a website built. It wasn't fancy, but it did the job, just like the old ad in the directory. Actually the two looked a lot alike. But Joe began to see his traffic and customer base diminish. This was because Internet searches for plumbers in his area didn't bring his website up first. In fact, his site didn't come up until page three of the search.

He did some research and found that he needed to add some things to his site to show up higher in the search engine. He came face-to-face with the new acronym: SEO. Search Engine Optimization became his key to getting his business's name out. So he started a blog and began tweaking the information on his site. But all this began to eat into the time he spent running a business.

Connections

At this point, the light bulb came on. Joe realized he was a plumber, not a writer. And this is where we come in. We can provide valuable services to businesses small and large. We can do what we do best—write—while allowing them to get on with business.

On with the Basics

The first term you need to become familiar with is Keywords.

A keyword is like a label. It's a short way—although almost always more than a single word in length—to state the purpose of your post. These can have several keyword groups or only one. If I wrote this piece for a blog post (and I did) my main keyword group for this article was Writing for the Internet. I also included Internet, Internet Audience, Learn the Basics of Writing for the Internet, How to use Keywords Effectively.

I used groups of words because the point of the keywords is to direct the searcher using Google to your website. You want your keywords to match, as closely as possible, what someone types into a search engine search box. People rarely type just one word because it gives too many options.

Here are the guidelines for using keywords effectively:

- Always use the keywords in the title.
- Repeat the keywords at least once in the first 50 words of your article.
- Spread the use of the keywords naturally and evenly throughout the rest of the article. (In a 400-word article that would mean using the keywords a minimum of three more times).

Anyone who's spent time writing content for the web, or even researching this market, has run across the acronym SEO. This acronym stands for Search Engine Optimization. It's basically where, in the list of millions, your content will show up when searched by a reader (search engine). This is determined by a closely guarded, mathematical equation called an algorithm.

When you use different search engines—Google, Yahoo, etc., you'll notice each will give slightly different results from any given search, because they each have a proprietary calculation. But there are things we can do as writers to move our content up in the rankings. To accomplish this we need a basic understanding of how SEO algorithms work.

Early on, these algorithms were less complex and depended heavily on keyword usage. Website writers would just use pages of keywords to raise their ranking. The search engines caught on and the equations got more complex. These algorithms will continue to evolve, with the goal of giving the searcher the most valuable sites first in the rankings.

Even with the evolution of algorithms, one thing hasn't changed. They are still mathematical equations and as such, Search Engine Algorithms take words literally—and that can be good or bad.

This means that they don't always understand double meanings or the clever usage of words. For example, a recipe for vegetarian chili cleverly titled, Too Hot to Handle Chili will rank far lower than one titled, Homemade Vegetarian Chili. This is because an algorithm uses the literal meaning of words and the first title doesn't even have the word "vegetarian" in it. Often times a clever title will result in fewer clicks.

This doesn't mean we can't be clever—only that we have to be deliberate in where we're clever. Take that chili recipe, give it a title that can be searched literally, like Hot and Spicy Vegetarian Chili, but in the description use the clever tag line as "too hot to handle."

Sometimes a clever title can work. My blog—**The Write Conversation**—is a clever play on words that works. I wanted the site to be searchable for the keyword "write" as well as be clever about educating writers as an ongoing conversation.

As a writer, I love to play with words. Part of that play is coming up with creative titles. Instead of seeing SEO as a hindrance to that, I look at it as an added level of difficulty in a game I enjoy more than any other.

This simple explanation should give any writer a good working knowledge of SEO. You can apply it to your own personal blog or website or you can use it to write effectively for clients.

Connections

Connect! Activity

Once you have your blog up and running, check out where it stands in different search engine rankings. It's important to know what comes up because this is exactly what a prospective editor or agent will do when they receive a submission from you.

Here are some things to investigate:

•   Search for your blog name in different search engines, including Google, Yahoo, Ask, and others.

•   Search for some of your blog post topics in these different search engines.

•   Search for the overall topic of your blog site. For example, my overall blog topic is writing instruction.

•   Finally, search for your name in these different search engines.

ANALYTICS

What the Numbers Really Mean

These days everyone's talking about analytics and how important they are. I'll give you the info you need to take your blog to the next level.

First let's start with the basics. **What are analytics?**

One way of defining analytics within the context of the Internet is this: a process of collection, measurement, and analysis of visitor activity on a specific website. This process is intended to help the author understand and achieve the goal of the website.

This definition brings up several important points. First you should have a goal for your blog. It's critical to your success that you know what you want to accomplish with it.

Do you want to:

* Connect with like-minded people?
* Share information?
* Build a platform or following?

All of these are great goals, but getting to each one will require a slightly different route. Before you can evaluate how well you're doing, you have to know where you're going.

Analytics can help you get there—if you know what you're looking for.

Connections

Here are the most important statistics to look for:

- Unique hits.
- Returning visitors.
- Traffic graph.

A unique hit is a statistic describing a unit of traffic to a website, counting each visitor only once within a specific time frame, say ten minutes. So visitor A could click on your blog once in a ten minute time period or 25 times during the same ten minute time period, and either way they would only be counted as one unique hit.

Returning visitors are important because we want people to come back regularly. This is the statistic that tracks this.

The Traffic Graph for your site helps you track the popularity of your posts and the timing of when people visit your site. This can give you valuable insight when you're planning your schedule and deciding what to blog about.

Connect! Activity

Take time to get acquainted with the analytics of your blog. I recommend you periodically check how numbers are doing and keep a record. Having this information nearby can help when you're putting together a proposal for an editor or agent.

LEARN TO TAG AND LABEL YOUR BLOG CORRECTLY

Even though successful tagging may seem like a complicated formula—one part magic and two parts luck—it's not as difficult as you may think. Tagging, sometimes called labeling, is closely related to keywords.

I used to be rather lackadaisical about tagging until I accidently stumbled on success—then I was hooked.

To understand how tagging can help your blog post get found (or lost in the sea of other blogs) you do have to understand a little about search engines. Search engines "crawl through" the Internet searching and sorting all the information. In the simplest of terms, the more your information about and on your page matches up, the higher your site ranks.

Now, for those of you who truly understand search engines, you know this is way over-simplifying the concept. Lots of other things come into count when your post is ranked, like the relevance of your post and how often you post, but tagging correctly will help increase your traffic.

Here are some of the basics of successful tagging:

• Start with your title. You want your title to contain your main keyword phrase.

• Repeat key words throughout the blog post. Don't overload your copy with words. Always make sure the writing is tight. Think of it as an exercise in focus!

Connections

- Re-use your title in the first couple of lines. It doesn't have to be an exact repeat of the phrase, just be sure to include the keywords.

- Tag with phrases, not just words. This is something a lot of people miss. When you tag (or label) your post, use your entire title. Then, if you have room, you can also include individual words. Think of it this way, when you enter something into Google's search box, do you enter just one word or a phrase?

I know this sounds like a lot of work, but I spend almost as much time on my title and keywords as I do on composing a blog post.

Below is an example blog post for you to see how I make the keyword phrase work for me. The main keyword phrase I chose was "Is Blogging the Right Choice for You?" It may seem like a catchier title would have been, "Is Blogging the Right Choice for Me," and that is the title I started out with. But it was almost impossible to work "ME" into the post naturally. This keyword phrase works well because it will pick up searches for, "Is blogging the right choice." The post is short, not even 300 words so I only used the keyword phrase twice in the body of the post. But it was important to use it in the first fifty words, as well as the title, and that it occurred naturally.

This next section is an example blog post. The other keyword phrases I chose for this post were, **Blogging and Blogging for Writers.**

Is Blogging the Right Choice for You?

For several years, blogging has been touted as the most efficient means of reaching an audience—especially for writers of all types. Now enough time has passed that it's possible to get a good statistical foundation of whether or not blogging is the right choice for you.

And the answer is . . . it depends.

I know, I can hear the groans from here, and I feel your pain. Everyone was promised, *"Blog and they will come."* Well, that is only partly true and here's the nitty-gritty of blogging.

Blogging works for writers under these circumstances:

- The blog/blogger has a clear purpose to blogging.

- The audience is clearly defined. (For example, a novelist is writing for writers—not readers.)

- The blog fits the picture of who the writer is. Or, in other words, it enhances—not contradicts—the author's brand.

- The rest of your branding makes sense with your blog.

Blogging DOES NOT work for writers under these circumstances:

- The point of the blog is vague and undefined.

- The audience isn't clearly defined.

- The blog leads readers to a different picture of the writer—not a deeper picture—but totally different.

Over the next chapters I will address each point and show you how to evaluate if blogging is the right choice for you. I'll help you make your blog work for you, instead of being a time waster for you and your reader. But for now, I'd like you to evaluate where you are with blogging.

- Do you have a blog?

- Is your blog's audience growing?

- Does it deliver results?

Connect! Activity

If you already have a blog and haven't been tagging or labeling your blog posts, it's not too late. You can raise the search engine rankings on those posts by going back in and adding tags and labels. Be sure to start with the most recent posts and work your way backward in time.

DON'T WASTE YOUR HYPERLINKS

A Hyperlink is a clickable link found within a post. Other common terms for a hyperlink are a **Hotlink** or just a **Link**. Most people don't know that the words hyperlinked in your post are also searchable by search engines. This means you need to choose the words you link with carefully, because you don't want to waste them.

Hyperlinks are helpful for several reasons:

- Including them will raise your search engine ranking.

- It's a way to utilize previous posts and get more traffic on your own blog.

- They can give your readers more value by linking to a related site that your readers may not have visited.

- It's a way to build credibility by linking to other sites you've written for.

Here are some tricks to formatting hyperlinks correctly to give you the best results.

- **Tie your hyperlink to valuable words.** The words hyperlinked in your post are also searchable by search engines and you don't want to waste them, so pick words that are specific to the topic you're linking. For example, people often link to the word *POST*. It makes sense when sending the reader to a previous post, but how often do people do a Google search for the word post? Not very often. So be sure to link to something people would type in a Google search window.

Connections

•   **Be sure the site you link to opens in a new window.** This is important because your website stays open and it's easy for your reader to check out the info you're referencing and return to your post. Otherwise you might lose them. In most standard blogs and websites you have an option to decide this.

•   **Always check your links.** Let me say that again—ALWAYS CHECK YOUR LINKS! Very few things will irritate people more than to be curious about something and then be directed to a dead-end when they try to learn more. I don't care how familiar I am with a link; I always test it before I publish a blog post.

•   **Inform others.** If you're linking to someone else's site or article, it's polite—and smart—to drop them a quick e-mail and let them know. We all like to know that others have found our information valuable. Frequently the honoree will tell others about your article and this can also help you get the word out about your blog or topic.

Connect! Activity

Just like the previous Connect! Activity, you can go back and correct any mistakes you have with hyperlinks. Be sure to start with the most recent posts and work your way backward in time.

Here's what to check for:

•   Make sure the hyperlink opens a new window instead of taking the reader completely away from your site.

•   Look for search-engine-friendly words to link to, things a reader would type into a search engine box.

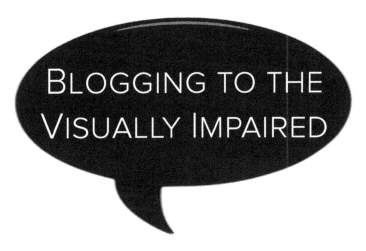

BLOGGING TO THE VISUALLY IMPAIRED

Before you begin to build your blog—or if you're updating it—a big consideration you need to include is how accessible your blog is to the visually impaired.

This may seem like an unusual recommendation, coming from someone with fairly normal sight, but it's one that's become critically important to me. I've discovered I'm missing an entire world of new connections if I don't take this into consideration.

Before last year I'd never considered that some blogs and websites were easier than others for the visually impaired to navigate. At that time I was blessed by an introduction to an amazing young woman (and talented blogger in her own right). The main differences between this young woman and me were her sweet spirit and gentle wisdom... and the fact that she's blind.

Jamie introduced me to the software she uses to navigate the world accessed through her computer. And the access is filled with hidden obstacles and obstructions that would have derailed me within moments. Her endurance and perseverance challenged me to reevaluate all the tools I used on my blog. I researched and investigated every stumbling block to see if it was a necessary obstacle or something I could remove.

What I found may surprise you, it certainly did me. For one, you don't have to be blind to be visually impaired. As our society ages, there are many of us who don't see as well as we used to and a lot of the measures we take to avoid spammers also block access to others. There's also a significant section of our population who struggle with color blindness to some degree or another. All of these people deserve equal access.

There are ways your site can easily facilitate that access. One of the most simple is keep the color scheme sparse. Black and white are the easiest to navigate, but that choice can cause eye strain for those without impairment. The idea is to keep your color choices strong, with good contrast so those who struggle can tell the difference between different sections. For example, the difference between regular text and a hyperlink within a post (a hyperlink is a clickable link that takes you to another webpage). The more dramatically different you can make each area, the more clarity your blog will have.

Special Note: Those with color blindness, (also called color vision confusion), frequently cannot differentiate between red and green, although yellow/blue confusion can also be an issue. Avoid these color combinations to remove unnecessary confusion.

Here is a website that will check your color choices for free. (**http://www. checkmycolours.com/**)

The Internet is evolving into a complex community that's breaking down walls. I believe it's important to make certain I'm not erecting barriers in my part of the neighborhood.

Connect! Activity

Here are some things to check to make certain you're doing everything you can to make your blog a welcoming place for everyone, no matter what their struggles.

- **Labels make life easier.** Labels or captions are important for every image. These should include an accurate description of the image. Also, when posting a list of websites, or anything else on the sidebar, be sure to label it. You should never rely on colors alone to convey information.

- **Use headers and introductory sentences for paragraphs.** Many readers used by the visually impaired have a scanning option. Users can listen to the header and first sentence and discover whether or not the paragraph is relevant to their search.

• **Make link destinations clear within the text.** This makes good sense at any time because hyperlinks are also searchable by search engines and can help raise your SEO (search engine optimization) for your blog.

• **Check your font size.** In Blogger, you can set your default font sizes for your posts in the template window. But most of the main blogging platforms such as Typepad and WordPress allow you to change your font size from the posting window. You can't set pixel size but you do have the options of smallest, small, normal, large and largest. Make sure your blog post can be read from at least three feet away in normal light.

• **Skip the Word Verification for comments when possible.** Depending on your platform this may not be an option, but I encourage you to use a platform where you do not have to use this. A lot of people have the incorrect assumption that this tool will stop spammers from posting horrible stuff in your comment section. In truth, if you use one of the big three platforms (Blogger, WordPress, Typepad) their spam detection software is more than adequate for this task.

PART THREE — SOCIAL NETWORKS:

FACEBOOK, TWITTER AND DEVELOPING RELATIONSHIPS

IT'S ALL ABOUT THE RELATIONSHIP

When I first got involved with social networking I had two goals:

1.   **Get my name out.**

2.   **Prove to the gatekeepers (editors/agents) I was publishable.**

I never considered interacting with other writers to be anything other than a side benefit. After all, writing is primarily a solitary pursuit, right? **Wrong.**

Writing for publication is an endeavor built on forging relationships. Those relationships can ultimately determine your success or failure in the writing industry. **Here's a list of those relationships:**

•   Between you and other writers.

•   Between you and the reader.

•   Between the reader and the subject or characters.

•   Between you and the editor.

•   Between you and your agent.

I listed the relationship between writers first, because surprisingly, it's often the most vital in your writing life. The actual act of putting words on paper is a solitary one and because of that, it's easy to lose perspective. Writing in a vacuum can give us a false sense of whether or not we're effective in our endeavor. We either wind up thinking we're a genius or sink into the depths of despair because we can't string two coherent sentences together. Rarely is either perspective accurate.

Connections

We need others in our profession to give us feedback, keep us grounded, and provide encouragement. Social networking is the perfect place to find these people. Let me add a word of caution here. You may be tempted, like I was at first, to insert friends and family into this role. Unless they're also writers, this dynamic just doesn't work. They'll unwittingly encourage you when you need a swift kick in the pants and administer the kick in the pants when you need encouragement.

Many of the most valuable relationships I've developed through the years have come from those I've met online. My mentors, my colleagues and yes, even many of my editors, have come from those relationships. That's an important fact to remember. People move around in this business, the beginning writer you befriend now may end up as your editor in ten years' time.

So what steps can you take to build these online relationships? **Here are my suggestions:**

- Answer comments on your blog/Facebook page/Twitter feed.

- Comment on other writers' sites.

- Offer something of value.

Many authors new to the social media world wonder how they can offer anything of value online. **Here are some ideas to help you get started:**

- **Promote someone else**. This gives you credibility with your audience. Don't be afraid of the competition. It may seem counterintuitive, but the relationships you build with others who offer similar products or services can advance both of you.

- **Celebrate another's success.** We all need to know success is possible. By sharing someone else's achievement, we encourage them as well as others who may wonder if anyone ever succeeds.

- **Don't waste people's time just to get your name out there.** Make certain that what you talk about online has value. (See the points above)

- **Keep it positive.** Don't ever forget that what you say online may outlast you. Take it from someone who knows—**don't say anything negative about a person, product or service**—ever.

- **Don't be a Social Media Hog**. Translated, this means keep yourself to no more than a couple of Facebook or Twitter updates in a row. Otherwise you'll highjack the social media stream.

These tips will help you begin relationships that will last throughout your writing journey.

Connect! Activity

Start now to build a library of quotes, questions, and comments that will help you engage others online. Here are some of the lists I suggest you make:

- Quotes that have to do with the focus of your blog.

- Open-ended questions that will promote dialogue.

- Graphics are huge on social media, so begin now to build a library of images you can use to start conversations.

SOCIAL MEDIA & DIMINISHING RETURNS

At the time of printing, there were approximately 40 additional social media platforms and networks. I say approximately because counting them is about as hard as numbering a charm (flock) of hummingbirds. And they come and go almost as quickly.

Trying to chase down and join all the networks is not only impossible, it's completely counterproductive. As writers, we need to remember the purpose of social networking, and it's not to make a career as a marketer. It's the means to an end—actually a couple of ends. We want to connect with our audience, as well as with other industry professionals.

To that end, there are three things I've found productive in almost 100 percent of situations. I call these the **Marketing Triad**.

- **Blogging**
- **Facebook**
- **Twitter**

In my opinion, those are the only three mandatory things a writer needs to do to have a valuable presence on the Internet. As you've read, I have a couple of other networks I enjoy hanging out on and find valuable, Google Plus and Pinterest. This doesn't mean those will be valuable for you. I have several friends who find LinkedIn valuable, and it has never been so for me.

Connections

I've found that social networking is governed by the law of diminishing returns. Spending more than 30 minutes a day (after you've come up to speed and are familiar with social networking tools and platforms) is not effective. And the longer you spend, the worse the numbers get.

Connect! Activity

I want you to make a pact with yourself right now. Promise yourself you won't obsess over the numbers. That kind of worry and micro-management will drive you to spend long hours on social networking. It won't lead to more page views, followers, or friends. Instead it only leads to burnout, frustration, and defeat.

Personal Profile or Professional Page?

Sometimes it seems that Facebook changes almost weekly. Nowadays there are almost as many options as there are Facebook users. But the question I hear the most from writers is:

"Should I have a personal page or a professional page?" Let's look at the difference between the two to help you answer this question.

Personal Profile

A personal profile is foundational for your Facebook membership. To have any other kind of page you must have a personal profile. (There is an exception if you open a business page, but those aren't really applicable for writers.)

Note: I've seen several posts lately on e-mail loops advocating opening multiple Facebook pages under different e-mail addresses/names, to manage family and work contacts separately. You should know this is violation of the Facebook User Agreement and will very likely get you banned from Facebook.

Connections

Why Stick With a Personal Page?

- It's probably where the majority of your friends are located.

- For some reason, many people will accept a friend offer, but won't "like" a page.

- It's personal and feels more like a friendship and less like a business relationship.

- You have less than 5000 friends. Although, I have over 600 friends and I'm finding that a personal page is becoming unwieldy to manage.

If you have more than 5000 friends, Facebook requires you to move to a Fan Page.

On a Fan Page you have the option to Pin a specific post to the top of your page for up to seven days. It's easy to do; simply select "edit" on a specific post and choose "pin to top." This way everyone who visits your page can see a free download or anything else you want to highlight. It's not just for those who visit for the first time.

Both types of Facebook pages now have a cover photo. There are some very specific guidelines in what you can do with your picture. Facebook is trying to get away from all the hard-sell marketing.

Timeline (personal profile pages) cover photos may NOT contain:

- Price or purchase information, such as 40% off or Download it at our website.

- Contact information, such as web address, e-mail, mailing address or other information intended for your page's "about" section.

- References to user interface elements, such as Like or Share, or any other Facebook site features.

- Calls to action, such as Get it now or Tell your Friends.

Privacy Concerns

If you haven't joined Facebook yet, I encourage you to take the plunge. There have been many warnings about Facebook and the way it violates privacy.

Facebook is a tool—not a good tool or a bad tool—just a tool. When used correctly it can be a great thing. It connects people and brings a lot of good. When used incorrectly, it can do great damage. But it isn't inherently good or bad.

That said, there are some privacy settings you probably should enable. You can find these under the Account Settings, under Privacy. You should limit all your posting visibility to **FRIENDS** only, not **FRIENDS OF FRIENDS**.

I keep my contact information public because I want to be found. As a writer, I want people to know my name.

I also keep all my **LOCATION** settings turned *off*. I can't think of any reason I'd want someone to know the location of where I'm posting from. If I do, I can mention it in the post.

With these privacy settings and a strong password, you should find yourself with a safe and manageable Facebook account.

Which to Choose

I wish I could give you a definitive answer about which to choose, personal profile or fan page, but I can't. Not because there isn't a good answer, but because the good answer keeps changing. Frustration with Facebook is so widespread it's almost a social media cliché.

I can say, at the time this book is being published, I'm throwing the majority of my energy to my personal profile. Facebook has made it difficult, if not impossible, to get interaction on a fan page without resorting to paid ads. The only thing I'm certain of with Facebook is that it will change... soon. As to whether that change will be good or bad, only time will tell.

Connections

Connect! Activity

Take a few minutes to look at the time you spend on Facebook.

- What catches your attention?

- What types of posts are you most likely to interact with?

Now some specifics. Look at some of the professional pages you visit most often on Facebook.

- What types of posts catch your attention there?

- Why do you visit a professional page? If your answer is you don't ever visit a professional page, then you probably have your answer at what's the best fit for you. It's not a page, but a profile.

Then look at the personal profiles you visit most.

- Again, what types of posts catch your attention there?

- Why do you visit someone's profile?

Answering these questions can give you a lot of insight into what works and what doesn't.

TIPS TO PROMOTE DIALOGUE

A lot of folks I've interviewed have complained that a professional page is all about promotion—almost a hard-sell mentality. This complaint stems from being unfamiliar with the medium. A professional page doesn't have to be about promoting yourself, and you don't have to sacrifice relationships to utilize a professional page.

How to Get People to LIKE a Professional (Fan) Page

• **Utilize a three-pronged approach to social media.** Use your blog, Twitter, and Facebook to reinforce each other. Set them up to cross post, and keep the conversation going.

• **Get on a schedule.** Set a goal for how often you want to post updates—then stick to it.

• **Give us the inside scoop.** Let your fans see you as a person—not just a business. (Warning: Don't go overboard with this. Just an occasional post about grandchildren or exercise goes a long way!)

• **Promote others on your Fan Page.** I know, it sounds counterintuitive but trust me, it works.

• **Visit the Facebook pages of your fans and friends.** You want people to comment on your page, so comment on theirs.

• **Put a Facebook button on your blog.** TWEET about your Facebook page and USE the appropriate HASHTAGS.

Connections

- **Decide what value you want to offer to your audience (fans).** Then deliver content related to that. Don't treat your Facebook Page like a bulletin board.

- **Utilize headline techniques when you compose your Facebook posts.**

- **Finally, stay up to date with your niche.** For example, if you write romance, give us an occasional scoop into the romance genre. Has a new publisher come on the market, an old one shut down?

Connect! Activity

The main idea is to get the conversation going on your page. The way Facebook has pages set up is harder, but it can be done. When you post on your page, make sure it's something that encourages conversation or is important to your audience. Remember, just because it's important to you, doesn't always mean it's valuable for them. Share things they can comment on, questions they can answer, lists they can add to, etc.

TWITTER CAN BE YOUR BEST SOCIAL MEDIA FRIEND

Twitter has all the community building capabilities of Facebook—in 140-character bursts.

However, it's an often misunderstood tool. New users take one look at the Twitter feed on the Twitter homepage and run screaming because it looks too chaotic to do any good. But the truth is, Twitter is a valuable weapon in your connections arsenal. It allows you to make meaningful contributions to a topic without hogging your time or the time of your followers. And, between Twitter and Facebook, you'll be able to reach about 95 percent of your audience. Without Twitter that statistic goes down to about 50–60 percent.

Think of Twitter like a short commercial spot during the Super Bowl—the place with the most viewers at one time. (See where I'm going here!) During those spots, some people are in the kitchen refilling their plate, others are chatting about the game, but a few are watching. Something in that particular commercial caught their attention. Those few spread the word and poof! a few become millions.

Twitter has this ability because it comes to us in short, 140-character bursts. These bursts are easy to share and reproduce. TWEETing, and reTWEETing—or sharing someone else's TWEETS—can cause a viral reaction. And that's a good thing. Having a TWEET that goes viral means your message is being seen by millions.

Connections

Setting up and Using a Twitter Account

Twitter is easy to join. Just visit the Twitter homepage and click on Get Started—Join. Set up your username and password. Click "I accept" then "Create my account."

At this point, Twitter will ask you if you want to find friends. Come back to this option later. Instead, check your settings.

Settings are very important on Twitter!

•   Enter your full name—only spammers and newbies hide their identities. You're on Twitter to be found.

•   Make sure the time zone field is set correctly.

•   Enter your location—again, you're here to be found.

•   Do not check Protect My Updates—this negates the whole point of TWEETing.

Say Cheese!

That's right—it's time to upload a photo. Photos are important. Spammers are known for having no photo. Personally, if you don't have a picture and I don't know you well, I won't follow you. (This is a common problem for those new to Twitter and can keep you from gaining followers.)

•   **First click on the picture tab.**

•   **Browse your files and select a photo.**

At this point you can click on the Design tab and do a little customization. There are several attractive options on Twitter.

Now that you've got your homepage set up, it's time to find some people to follow. You can import contacts from Gmail, Yahoo, AOL, etc. but you don't have to. Many people dislike invitations to follow someone and prefer to choose whom they follow.

I recommend you take your time here. Instead of inviting everyone you know, search for some close friends/associates that you'd like to follow. Start with about 20 to get accustomed to the feeds.

A note on etiquette here—if someone follows you, it's considered common courtesy to follow them back. This etiquette isn't set in stone. If you're really not interested, don't bother. But unless there's a good reason not to follow them, you should.

Common Terms

When you post something it's called Updating Your Status or TWEETing. Michael Hyatt has the best explanation of the Twitter community I've ever read. He likens it to sitting in a roomful of people.

- Updating your status tells everyone in the room something.

- Replying to an update answers a specific person, but everyone can hear (all your followers can see your update). You do this with the @ symbol, followed by the person's name. Example @EdieMelson. This is actually a clickable link so others can click on @EdieMelson to get to their feed.

- A Direct Message is like whispering in someone's ear. The message goes to them only.

- ReTWEET is when you repost or repeat something someone else has TWEETed.

Composing an Effective TWEET

Now that you're signed up with Twitter, it's time to start using it. The point of Twitter is to engage others online. **Here is the anatomy of effective TWEETing:**

- **First, determine what your goal is with the TWEET.** Do you want to engage people in a short conversation, send them to a website, or celebrate someone else? Depending on the goal, your TWEET will be set up a little differently.

- **Engaging in a conversation.** This means you want people to respond to your TWEET by answering a question, or giving an opinion. Here's an example of a TWEET for that: **What is your biggest pet peeve on #Twitter?** Unless I'm referencing a website or a person, there's no need for a link or a person's name. I used the HASHTAG for Twitter (#Twitter) to make the TWEET searchable and hopefully engage some new people.

Connections

- **Sending someone to a website.** Here's an example for a TWEET for that: **Why I've switched from TWEETdeck to Hootsuite! http://ow.ly/fkcQL #socialmedia.** In this TWEET I have a link to a blog post I wrote. I have shortened the link (using Hootsuite) because it would take up too many of my 140 characters. I've added the HASHTAG for social media to again engage new people.

- **Celebrate someone else.** This could be by giving them credit for a blog post, or congratulating them on a book contract. Here's an example for a TWEET like that: **Why is your second novel so important? from @chipmacgregor http://ow.ly/fjs0h #write.** In this TWEET I've used the at (@) sign to mention someone's name.

Connect! Activity

Spend some time right now either signing up for Twitter, or tweaking your personal information. Do it while it's fresh in your mind.

USING TWITTER IN THE MARKETING TRIAD

There are certain things that we can do to help others find our valuable TWEETS. Let's go back to the Super Bowl analogy for a moment. During the game, some of us take watching these commercials a step further—we're watching for certain ads—think of it as looking for keywords that pertain to our wants or needs. TWITTER is set up to accomplish the same thing, through the use of HASHTAGS (#) and the use of @.

HASHTAGS or # denote a subject. People can search TWITTER for certain subjects, like writing or publishers. So if I want a certain TWEET to reach other writers who don't yet follow me, I can insert #write somewhere within my TWEET. I started using HASHTAGS in my TWEETS and within TWO HOURS added 26 new followers.

A word of caution, don't overuse the HASHTAG. If you use it too often within one post it will seem like a hard sell and you only have 140 characters to get your message out.

As a TWITTER user, I search for these subjects and have them arranged in LISTS. When I find one that's interesting, I can click on it to investigate further and then FOLLOW that person.

@ designates a person or a company. Just like I can search for a subject, I can search for a specific entity. I follow one geeky, technical blog called MASHABLE. I love their posts because I can understand them and apply them. Their TWEETS are even better, so I created a list that follows every mention of MASHABLE.

Connections

Twitter is a versatile medium, but because of its condensed format TWEETS that are about me, me, me get really old, really fast! Sure we all want to increase our audience and reach more people, but there are better ways to accomplish that. The thing you want to be known for on Twitter is relevant content. You want to give those who follow you something worth following.

Another helpful tool is—Lists. This is simply pulling together TWEETS of one particular subject and allowing others—those who have joined your list—to follow them as well.

To make up your own list, click on "List" on your Twitter homepage and follow the directions.

Connect! Activity

Now is the time to play around with HASHTAGS. Visit your Twitter homepage and type different HASHTAGS into the search box at the top and look through the TWEETS that come up. You'll get some insight into what that HASHTAG represents, and you'll probably find some pretty interesting people to connect with.

MAKE EVERY TWEET COUNT

More and more folks these days are discovering Twitter. That's a good thing because it's easier to connect with more people.

It's also bad, because the population explosion makes it harder to stand out.

Here are my tips to make every TWEET count!

- Let someone else benefit from what you've found helpful.
- Share about someone else's success.
- Pass along opportunities.

TWEETing, reTWEETing, and posting are perfect ways to do this.

When you compose an announcement for Facebook or Twitter, keep your audience in mind and use these ideas to whet their curiosity:

- Use open-ended questions.
- Bring up an intriguing point.
- Hit on a subject that everyone struggles with and show that you have an answer.

Don't make these mistakes:

- Don't give away the ending.
- Don't sum up your post.
- Don't give them a reason to not visit your site.

Connections

And the most important Twitter Tip:

Remember, it's a conversation—not a platform for self-advertising!

Connect! Activity

It's time for another list or two. This time I want you to come up with some ideas of ways you can connect with others within the 140-character burst of Twitter.

- Start a list of thought-provoking quotes that have to do with you.

- Make a list of industry professionals that you can recommend others follow.

- Come up with some open-ended questions you could TWEET that would engage conversation.

TWITTER ETIQUETTE

Which Should I Do, Reply or ReTWEET?

A REPLY is used when you answer a comment, or thank someone for mentioning you.

It's good form to reply to someone who reTWEETS something you said.

A reTWEET is used when you want to repeat someone else's TWEET.

It's poor form to reTWEET a TWEET that mentions you. It's not an unforgivable error, we've all accidentally hit the reTWEET button instead of REPLY, just try not to make it a habit.

Is There a Rule about How Often I Should or Shouldn't TWEET?

Twitter has a 140-character limit for a reason and it's considered bad form to TWEET 3 or more times right in a row. Doing this is called Hogging the Stream or Hijacking the Stream. When I schedule my TWEETS, I try not to schedule them for closer than ten-minute increments.

What's Up With All Those @ Lists People TWEET?

There are several times you may see this.

First, it's polite to thank new followers, and some people group them together in lists. This isn't a bad thing, if you don't hog the Twitter stream doing it.

Connections

Second, there are certain designations for days of the week. For example, Friday is often designated as #FF. This stands for FOLLOW FRIDAY. It's a time for people to recommend others as valuable people to follow. I've actually found some really cool folks to follow through this. And once again, it's not a bad thing, if the person doing it doesn't Hijack the Twitter stream with multiple #FF posts in a row.

Do I have to Follow Everyone Who Follows Me?

No, absolutely not. I try to follow the people I believe have something valuable to say. I don't follow those who look like spammers or sketchy accounts. Often times you can tell because they don't have a picture, only that irritating egg avatar. Another way I check is to look at the description they give for themselves. If they don't have a description, I'm always skeptical.

What Does RT, MM, OH, HT, and WW Mean?

RT—means RETWEET.

MM—is MUSIC MONDAY, similar to FOLLOW FRIDAY.

OH—is OVERHEARD.

HT—is HEARD THROUGH.

WW—is WORDLESS WEDNESDAY

Connect! Activity

Take a minute and look through your list of Twitter followers. To find them, visit your Twitter homepage and click on Followers. Make sure you've followed everyone back you want to.

SIX TWITTER TABOOS

Number 1—always talking about yourself. With the 140-character limit, TWEETS that are always me, me, me, are guaranteed to not only keep you from new Twitter followers, it will help you lose the ones you have.

Number 2—hijacking the Twitter feed. This means posting three or more TWEETS in a row. Again, it's that nausea-inducing litany of me, me, me.

Number 3—loading your TWEETS with HASHTAGS. Multiple HASHTAGS are sure signs of spammers, bots, or newbies. All accounts no one wants to follow

Number 4—a missing picture or the dreaded egg avatar. Again, only spammers, bots, or newbies don't change out the egg for a real picture. You only get the opportunity to communicate who you are in a 140-character bursts. Trust me, lacking a picture does NOT inspire me to follow you.

Number 5—a locked Twitter account. Twitter is all about being found. A locked Twitter account is the ultimate oxymoron. Some people restrict who follows them because they mistakenly believe their followers' TWEETS will end up in their Twitter stream. Not true. Only the people YOU follow show up in your Twitter stream.

Number 6—a description that promises you'll be commenting on the random or inconsequential. Twitter is all about focused bursts of info and interaction. Keywords and HASHTAGS help us follow specific topics. Random nothingness is not what I'm looking for on Twitter.

Connections

Connect! Activity

Visit your Twitter homepage and check on the things mentioned above. Spend some time composing a description that conveys who you are. If you aren't sure what to put, visit some of the people you follow and look at their description.

FINDING RELEVANT HASHTAGS

It's important to use HASHTAGS when you TWEET—but it's even more important to use them correctly. One big thing you should avoid is using more than two or three HASHTAGS per TWEET. Otherwise you run the risk of being mistaken for a sales person or a spammer.

If you're TWEETing about a new subject, be sure to check the HASHTAGS and pick ones that will correctly target your audience. For example, I'm targeting military families with my book, *Fighting Fear: Winning the War at Home When Your Soldier Leaves for Battle.*

Since I don't normally receive TWEETS about military subjects, I began to research the best HASHTAGS for my targeted audience. The first one I chose was #military. It seemed like a good choice, but when I researched it, I saw a lot of TWEETS about the best times to meet soldiers to get a date. Definitely not the audience I was looking for. Next I tried #militaryfamily and that took me to TWEETS directed at the people I wanted to reach.

I know your next question is going to be: "Where does someone research HASHTAGS?"

There are three sites I recommend. All three are free resources and I don't recommend one over the others, because they're all slightly different and I find myself using them in different combinations.

The first is **HASHTAGS.org**. This site gives a list of trending HASHTAGS, a graph, and a search box to type in potential HASHTAGS. After typing in a HASHTAG, it lists current TWEETS containing that HASHTAG.

Connections

Another is **WhatTheTrend.com**. Like the others, What the Trend has current trends, a graph, and a search box. I included it because it also explains why a particular HASHTAG is trending.

Finally, visit **Trendsmap.com** regularly. This site has a map that tells you what is trending where. This is particularly helpful if you're targeting a specific area.

Connect! Activity

Spend some time right now researching HASHTAGS that relate to you. Then write them down. Keep them someplace easy to access so you'll have them when you need them.

HEADLINE TECHNIQUES

The dictionary defines a copywriter as one who "writes copy for advertising." The field of copywriting has gone on to include many aspects of business writing, and it has particular applications for Twitter and composing a powerful headline.

The primary goal for all copywriters when they compose a headline is not, as you might assume, to sell something. **It's to get the first sentence read.**

That's where the similarity with social networking begins.

Your choice of Headline, Graphics Font, Format, etc., should lead directly to this goal. And what is their goal with the first sentence? To get the next sentence read.

This step-by-step copywriting road is also the yellow brick road for everyone who wants to succeed with Social Media.

You see, with a compelling headline, a browser becomes a reader. Without a compelling headline, the rest of your words might as well not be written. The same could be said of your Facebook post or Twitter update.

Connections

So what makes a compelling headline/social media update?

- It must provide the reader with the tools to evaluate the content.

- It needs to resonate with a reader's urgency.

- It's important to show the reader why this offer/product/person is unique.

- And most important of all, it must not give away the ending!

Not sure what that means?

Let's say I'd written a blog post about how Hootsuite can help you organize your Twitter life.

Here's the TWEET/Facebook update I'd send out:

Does the confusion of Twitter make you crazy? Visit my blog today for 5 tips to organize your TWEET Life.

Here's a TWEET/Facebook update I'd never send out:

Does the confusion of Twitter make you crazy? Use Hootsuite to organize your TWEET life.

With the second one, I've given away the ending and there's no reason for anyone to visit my blog.

Now, before we get too much deeper, I want to give you a list of all the different headlines you can create.

Direct Headlines go straight to the heart of the matter, without any attempt at cleverness. A direct blog post title might read **Free SEO E-book**.

An **Indirect Headline** takes a more subtle approach. It uses curiosity to raise a question in the reader's mind, which the body copy answers. Often a double meaning is utilized, which is useful online. A blog post I once saw was **7 Reasons NOT to Blog**.

A **News Headline** is pretty self-explanatory, as long as the news itself is actually, well... news. A product announcement, an improved version, or even a content scoop can be the basis of a compelling news headline. **Think Introducing the New Google Plus**.

The **How-to Headline** is popular everywhere, online and off, for one reason only—it works. An example would be, umm... oh yes... the title of this chapter.

A **Question Headline** must do more than simply ask a question, it must be a question that resonates with the reader. One used far too often in the writing industry is **Do You Want to Get Published Now**?

The **Command Headline** emphatically tells the reader what to do. The first word should be a strong verb demanding action, such as **Subscribe to The Write Conversation Today!**

Another effective technique is called the **Reason Why Headline**. Your body text consists of a numbered list of product features or tips, which you then incorporate into the headline, such as **8 Ways to Build a Platform**.

Finally, we have the **Testimonial Headline**, which is highly effective because it presents outside proof that you offer great value. This entails taking what someone else has said about you, your product or service, and using their actual words in your headline. Quotation marks let the reader know that they are reading a testimonial, which will continue in the body copy.

Remember, it's important to make sure you vary the type of headline post you use. In other words, don't always use a "question" or a "reason why" headline. **Shake it up a little.**

Connect! Activity

Take an old blog post and come up with different headlines for it. Get used to composing several social media updates when you write a post. This will save you time when you're advertising your blog on your social media profiles.

Part Four — Social Networks:

Afterthought Platforms that Shouldn't be Afterthoughts

YouTube — It's Big

YouTube is the second-largest search engine in the world, and it's owned by Google. These days, the difference between the search engine rank of a website with videos and one without is considerable. Videos carry a lot of weight with Google's search engine algorithm. Videos are also much more likely to go viral than text-only articles.

A lot of people go to YouTube directly to search for info-movies. This form of information is the video form of an article. Although presently there isn't as much competition for video rankings in YouTube as there is for article rankings with Google, the Internet world is headed that direction at a breakneck speed.

Since the competition for YouTube ranking isn't as great, it's going to be easier to establish a ranking now than wait and try to catch up. **Here are some things you need to consider as you lay the groundwork for a solid presence on YouTube.**

Keywords Still Rule. It's critical to use your main keyword first (or as early as possible) in the video's title, description, and tags. You want to also use your own name, brand, or website as well, but that should come after the keyword.

The Title of Your Video is Critical. It's important that the title is a good representation of the information the video contains. Although it may seem too basic to mention, you'd be amazed at the number of people who still title their files "untitled.mov" or something else generic.

Connections

Include Your Keywords in the Video Voiceover. We're all becoming more proficient at SEO for articles. Voiceovers are spoken articles and must be just as carefully keyworded for SEO.

It's important to note that Google has a speech-to-text conversion technology which can convert your info-movie's voiceover into captions. Google uses this to improve video search technology.

Upload a Transcript File for Video Captioning. Uploading your own caption transcript is a much better choice than letting YouTube transcribe the audio. By uploading your own material, you have total control over what appears in your video captions.

To upload a transcript file, click on Edit Video in the YouTube Video Manager. Click on the Captions tab. Under the Add New Captions or Transcript header, select Transcript File as the Type and upload your script file.

To watch YouTube set up your file, click the CC button on the video and YouTube will display the words in exact sync with the audio.

Connect! Activity

Here are some things you can do to start your YouTube presence out right:

• **Build a Valuable YouTube Channel.** Start branding your YouTube Channel the same way you would a new website or blog. If you've already begun to experiment with videos, go back and apply the suggestions.

• **Add New Videos to Your Channel Regularly.** This is just like blogging, regular, new content gives you much more power in the search engines.

• **Respond to Comments on Your Videos.** Again, just as with blogging, answering comments is vital.

• **Look for Ways to Exchange Links with Other Sites.** Just like blogs and websites, it's important to prove your Channel is valuable to others.

Do You Pinterest?

Pinterest is growing by leaps and bounds. Although it's not one of my must use sites for writers, it's worth your time to check out. Here are some of the ways writers can utilize Pinterest:

Applications:

- Build boards with your book's subject.

- Build boards that help you with world building or setting.

- Share blogs from others who write in your genre.

- Build boards from books you love—you can legally post the book covers and links to Amazon or the author's blog.

- Build a board and share the blogs you follow—as long as the blog has a picture you can post it.

- Build boards with your personal interests, like knitting, sewing, or bird watching. There's no limit to the topics you can choose.

Other important tips:

- **You can set your cover for any board you set up.** This is the image that comes up on any particular board. Go to a specific board and hover over the pins, you will see a pop-up option of SET BOARD COVER. **Click this to choose the image as your board cover.**

- **NOTE:** if you've been added as a pinner to a board, you can't change the cover photo.

Connections

- **It doesn't matter who follows you.** You don't have the option to approve followers. If someone is doing something inappropriate, you can report him/her to Pinterest.

- **You can change the arrangement of your boards.** Go to your home page and in the center of the bar above your boards, you'll see the words EDIT PROFILE. There is a small icon to the right of that. Click the icon to rearrange the order of your boards.

Pinterest Etiquette

It's common to follow the boards of those who follow you, but you are under no obligation to. Like all social media, it's important to avoid the hard-sell approach. If your boards are all about you, it's advertising, not sharing.

Only pin images that are either copyright free, owned by you, or are okay to be pinned. **How to tell?**

- Look for a PIN IT link on the site you're on.

- Or the site says it's okay.

- Or, ask if you have permission to do this.

Connect! Activity

Here are some valuable boards for an author to have:

Story-building—This could consist of possible settings, visual representations of characters, historical information, etc.

Your personal hobbies—Readers are interested in authors. For some reason they often see us as mysterious and unapproachable. Posting about things that interest you helps them connect.

Recipes that your book characters would like or you have mentioned in your books—Recipes are a huge thing on Pinterest, and you should take advantage of that.

PINTEREST AND COPYRIGHT ISSUES, A WARNING

Pinterest is a great resource for writers. I love my Pinterest boards and have no plans to stop using this site. But lately some concerns have surfaced regarding copyright issues. This short article will help you navigate the murky waters of what's legal to Pin and Post on the Internet.

- **Just Because It's Available on the Internet Does NOT Mean You Can Use It**—All graphics and images are covered by an automatic copyright—just like our words. The image doesn't have to have a copyright symbol (c) for it to be protected.

- **Free Images are not the Same as Copyright Free Images**—Again, be sure to check the guidelines on a website before borrowing an image. Some people are fine with this if you just give them credit, others are not.

- **Linking Back to the Original Site does not Give You Permission to Use It**—Many people believe that if they insert a link to the site where the image originated, it's okay to use it. **This isn't true.**

It's also important to note that if you violate a copyright with one of your Pinterest boards you, and you alone, are liable for any fines or charges. You agreed to this when you opened your Pinterest account and accepted their terms of use. If you want to read them again, here is the direct link: **http://pinterest.com/about/terms/**

Connections

Additional Thoughts

I've also had friends who, in researching where an image originated, discovered that it was a cropped section of an image that was inappropriate in content. For me, this just reinforced the fact that I needed to be diligent in tracking down the source of what I'm promoting on the Internet.

Exceptions

There are a few exceptions to the copyright issue and those can be found mainly on Facebook and other networking sites. With Facebook, if you post an image, you are agreeing to its public use.

You still need to be careful, and make certain that the image you want to use wasn't posted illegally by someone else who didn't hold the original copyright. But, for example, if I posted a picture of Niagara Falls from my vacation on Facebook, you can legally use that picture without permission. By posting it on Facebook, I gave that permission away.

Bottom Line

It only takes a moment to send an e-mail and ask if an image may be used. We would expect nothing less if someone wanted to use something we wrote. We also need to consider our reputation as Christian writers. I feel that if I use that designation, I must hold myself to a standard that Jesus would approve of. I often fall short, but I'm constantly striving to represent Him well.

Connect! Activity

Take a moment to check those images on your Pinterest boards. Clean off the ones you're unsure of and move forward with a clean slate.

GOOGLE PLUS

Google Plus is relatively new to the social media world, but it's growing daily, in size and in value. It's a network that offers an interesting cross between Twitter and Facebook. And, as the name implies, it's owned by the Internet giant, Google.

Google Plus has given us more control over how we classify our networks through something called circles. We don't simply have to friend someone or follow them. With this network we can decide when, where, and how we interact with them.

For me this difference is key, and it's intuitive for me because I already group the people I interact with into categories. True, some fit in multiple categories, but even that is possible with Google Plus Circles.

Also, in the Google search engine, everything you post has extra SEO (Search Engine Optimization—how high your post ranks when someone searches for that topic or for you.)

Another thing I really love about Google Plus is the Hangouts. It's a place where you can video chat with up to ten people at once—and it's free. You can even record and broadcast from this feature.

Connections

Here's a quick note about circles. Google Plus comes with several default circles:

- Friends.

- Family.

- Acquaintances.

- Following.

I started with these defaults, but almost immediately began branching out. Now I have fun with it. I have a circle for besties (a complete girl term), buds, and even terminally boring. And the greatest part is no one knows the category I've put them in.

Please feel free to add me to your circles! You can find me under Edie Melson.

So what about Google Plus professional pages?

Google Plus gives you the option to have a personal page and a professional page. I recommend you begin with a personal page and then move to a professional page.

The big difference between professional pages and a personal page is the fact that you can't add anyone to your circles on your professional page until they have added (followed) your page.

Like Twitter, you need to follow other pages and if they find you relevant, they'll follow you back. Then you can add them to your circles on your page. Other than that, page circles work just like the circles on your Google Plus profile.

One of the biggest complaints I hear about social networks is confusion about privacy settings. Frequently this confusion can leave users exposed without even knowing it. Google Plus gives the user the opportunity to customize many settings in a straight forward manner and bypass the confusion.

In the privacy settings page, the first tab you want to consider is your visibility to search engines. On the right side of the screen you'll see a light gray box with EDIT VISIBILITY ON PROFILE written in it. Click on this and it will take you to your profile page.

At the bottom of the page you'll see, PROFILE DISCOVERY, after that, you'll see one of two things:

- PROFILE VISIBLE IN SEARCH
- PROFILE NOT VISIBLE IN SEARCH

For writers, I recommend you make certain your profile is visible in search.

Privacy and Circles

With the circles in Google Plus you have the option of different circles, family, friends, acquaintances, etc. There may be times when it's appropriate to share things with family or close friends that you don't want shared with business associates or acquaintances. You manage that setting every time you post information (a picture, link, post, etc). This may seem like a lot of work, but it's actually a really good failsafe for your privacy. Each time you post, you can decide whom you want to see what you've written.

SPECIAL NOTE: There is an automatic default with this setting. When you post something to your stream (your feed) and you designate who can see the post, Google Plus will remember this setting and default to it when you post again unless you change it.

Unless you're a celebrity, I recommend you do not make yourself invisible. Showing up in someone else's profile is a great way to network and make new contacts.

SPECIAL NOTE: Remember, with Google Plus only you know the specific names of your circles and whom you've added to each one.

There are other settings you can tweak on this page, but these are the major settings you need to be concerned with.

Connect! Activity

Spend some time on Google Plus. Get to know all the nifty features it offers, from tweaking your personal profile, to utilizing Hang Outs. Becoming familiar will help you connect more efficiently through this social media outlet.

KLOUT

First, let me assure you Klout is a legitimate tool, and not spam. Responding to a Klout invitation will not get you hacked.

What is Klout?

Klout is a measurement tool to gage your success in social media. It isn't actually a network, although you do interact with others there. Rather, it enables you to track your own influence in social media and connect with others and become an influencer. It does not track stats for your blog or website.

I pay attention to my Klout Score because it helps me stay consistent with my social networking. Beyond that, it gives others an unbiased snapshot of my reach or influence within the world of social media. This is important if you're trying to show a potential publisher you have a platform.

You may wonder how Klout does this. Klout correlates your social media presence by looking at Twitter—how often you TWEET, how many reTWEETS, as well as how many mentions you have. It looks at Facebook—your likes, your posts, how often you comment and how often others comment on your posts. It also looks at Google Plus, search engines and your websites and blogs. And it's constantly updating its resources and adding new data. So be sure to give it access to your accounts so it can give an accurate snapshot.

If I've piqued your interest and you'd like to sign up, here are some of the basics and an explanation of the terminology.

Connections

It's important to not get obsessed with the actual number of your Klout score. The average is around 30. Depending on what your goal is, your number may be smaller or larger. I'm fairly satisfied with my number, it shows I have a good influence and am very active in social media. Only high-level marketers and such tend to have scores in the 80s and 90s. You mainly want to look for consistency, with slow growth.

Sign Up for Klout

To sign up, go to **www.klout.com.** Don't be afraid to sign in through Twitter or Facebook. You'll need to give your permission for Klout to access both networks to accurately track your score. If you don't want to do this, I'd skip Klout. Without access to your Facebook and Twitter accounts, Klout can't give you accurate information.

Once you sign up, you'll be asked to choose categories that interest you and then name influencers within those categories. Think of this as a baseline for Klout to begin getting to know you. Be honest, but don't stress out about this section. It's just a starting point.

During the sign-up process and afterward, Klout will frequently give you the opportunity to TWEET or Update your Facebook profile with things you've done on Klout. This is totally up to you. You won't be penalized for not sharing; it's just a way to let people know you're part of this network and for Klout to get their name out. Sometimes I allow the info to go out and sometimes I don't.

Basic Klout Terms

Once you're a member of Klout, you'll have the opportunity to give a +K or receive a +K. This is similar to a mention on Twitter. If someone gives you a +K, they are saying you have been an influential source for a specific topic. (I've gotten several +K's for blogging.)

It's considered good etiquette to thank someone publicly, either on Twitter or Facebook. It's also common practice to return the compliment by giving them a +K as well. I was unaware of this and didn't do it the first couple of times and have been trying to correct my inadvertent faux pas.

Connect! Activity

If you haven't taken the Klout plunge yet, use this time to play with it. Go ahead and sign up for an account and see if what pops up is interesting and helpful or just frustrating. You'll never know until you try.

Part Five — Knowledge & Tools:

Bringing it All Together

COPYRIGHT 101

Take Time to Learn What You Can and Cannot Post on Your Blog

Lately, I've seen a lot of discussion about copyright issues. Specifically, what is legal to use on a blog, website, or in a book. It's critical that each of us has a working knowledge of what we can and cannot use on our blogs legally.

That said, I'm not a lawyer and this should not be construed as legal advice. If you've been notified that you are in violation of copyright, I recommend you contact legal counsel.

Instead of giving you just the facts, I thought we'd have some fun, and take a little quiz.

COPYRIGHT QUIZ—HOW FAMILIAR ARE YOU WITH THE LAW?

All the answers are True or False, so let's get on with the quiz!

1. I can legally post any picture on my blog if I link back to the place I got it.

FALSE: photos, sketches, and graphics are covered by the same copyright law as our written words.

Unfortunately, there is lots of sharing going on over the Internet, and it's not all legal. When we borrow photos without permission, even when we acknowledge where we got it, we are stealing. I truly believe that's not the intent, but we need to educate ourselves on what's right and what's not and then lead by example.

Connections

2. If I can find it in public places, like the World Wide Web, it's in public domain and legal to use.

FALSE: Public domain is a legal definition and has nothing to do with the availability of an image or text or music. For instance, do you know why many times people don't use the familiar Happy Birthday song? It's because someone owns the copyright to it and, except for private use, it's illegal to use without permission.

This is particularly true of images you find on the Internet. I can't count the number of sites I've visited where there are watermarked pictures everywhere.

3. I can legally use a song's title in a post, article, or book.

TRUE: Song titles are the ONLY part of a song we may use legally. There is something in the copyright law called fair use. Without boring you by quoting the law, it means that you can refer to part of a work without being sued. Contrary to what some think, there is no set number of words or percentage that makes up fair use. **Instead, there are four factors used to define it:**

- The purpose and character of your use.

- The nature of the copyrighted work.

- What amount and proportion of the whole work was taken.

- The effect of the use upon the potential market for or value of the copyrighted work.

Because of the small size of a song, say compared to a book, the courts have decided that quoting any portion of a song, except the title is a copyright infringement. The reason you can quote a title? Titles cannot be copyrighted.

4. I can legally use someone's blog post as long as I give them credit and don't change anything.

FALSE: You cannot legally borrow someone's blog post without their permission, even if you give them credit or link back to it. Now, there are some sites that post guidelines for you to follow to be able to use their posts. But this doesn't mean it's legal to do so for another site. The only exception to this is the Reblog button found on some blogs. This allows someone to repost the blog article legally.

5. I can legally use music or a song as background for an original video as long as I credit the source.

FALSE: I imagine many of you got this one correct. There has been so much written and so many fines levied in regard to stolen music, it's almost common knowledge that you cannot borrow a song or music for your own purposes.

6. I can legally post YouTube videos on my blog or website.

TRUE: This one was a little tricky. You can legally post YouTube videos on your site, because what you're doing is linking, not reposting. Even when you embed videos, they are still linked to YouTube.

7. If I don't make money off of it, it's legal for me to use.

FALSE: Whether you profit from borrowing someone else's work has no bearing on the legality.

I have an opinion about all the borrowing that is happening around the Internet. I may be an optimist, but this is my personal opinion.

I think a lot of bloggers are generous folks... to a fault. They frequently offer their own work to others for free. With this mindset as a foundation, it doesn't always occur to them to think of what they're doing as stealing—they don't see other people's "borrowing" as stealing, after all.

That said, I applaud the generosity with our own work. But, we should also be willing to guard the uniqueness and value of the work of others.

8. I can legally quote a small percentage of the words to a song in a post, article, or book.

FALSE: We are all used to being able to quote passages from books and not get into any kind of copyright infringement. That is NOT the case with a song. As covered above, the only part of a song you may quote is the title.

If you've seen songs quoted in published books either someone paid a use fee or the author wrote the song him- or herself. I've known of two authors who self-published books and had to pull the books because of songs quoted without permission.

Connections

9. If I don't have a copyright symbol on my work, it's not covered by copyright law.

FALSE: Copyright symbols are visual REMINDERS that what you're reading belongs to someone. Just because there isn't one, doesn't affect the status of what you see in print or online. If someone wrote it, it's copyrighted.

SPECIAL NOTE: You do not have to apply for a copyright for your work... **EVER**. You can register your copyright, but it's expensive and cumbersome to do. And it's rarely necessary.

10. I can legally download photos from Facebook to use on my own site.

TRUE: This is another trick question I snuck in. Because of the user agreement you signed when you registered for a Facebook account you agreed that your photos were able to be used by them for different online purposes. This makes it very difficult to prove in a court of law that you don't mean that permission for everyone else on Facebook. So, if I post my Niagara Falls vacation photos on Facebook, I can't complain if you borrow them.

Now, it's always good manners to ask permission, but it's probably not going to get you in any legal trouble if you use them without it.

ANOTHER SPECIAL NOTE: If the person posting the photo did so illegally, and you repost it, then you are just as guilty and can also be charged with copyright infringement.

11. I can legally pin anything to one of my Pinterest Boards.

FALSE: I don't know about you, but I love Pinterest! I guess I'm just a visual kind of girl. But there are a lot of folks getting into trouble on Pinterest right now. We have to follow ALL the copyright rules when we're pinning, just like when we're posting on our blogs. And, if you violate a copyright with one of your Pinterest boards you, and you alone, are liable for any fines or charges. You agreed to this when you opened your Pinterest account and accepted their terms of use. If you want to read them again, here is the direct link: **http://pinterest.com/about/terms/**

But there is one slight loophole. If someone or some business has a Pinterest button on their website, you can assume they want their stuff to be pinned and you should be okay.

12. I can legally post a picture of a book cover I recommend or am reviewing.

TRUE: This is an instance of Fair Use. As long as you're not saying the book in question is written by you (if it's not), you can legally post a review and use the cover.

13. Copyright on written works expires 70 years after it was first published.

FALSE: A lot of folks have heard that copyrights expire after 70 years. In some cases that's true... but not all. There are some instances when copyright expires 70 years after the author's/creator's death. There are also times when copyrights are renewed. Beyond that, there are other exceptions, so while the 70-year rule is a good place to start—it's not the place to end.

This chapter was not generated to scare you, but to give you confidence in what you're doing and doing well. Being a writer would give me a reason to be passionate about this, but I also come from a creative family. My mother is an internationally known artist and my dad is a classical musician, as well as a freelance photographer. Believe me when I say I cut my teeth on this stuff. Back in the day, I've known my mother to correct complete strangers standing in front of copying machines with art books.

Resources·

Good explanations of copyright:

- http://www.copyright.gov/laws/
- http://www.rbs2.com/copyr.htm
- http://www.bitlaw.com/copyright/index.html
- http://library.findlaw.com/1999/Jan/1/241476.html

Public Domain Info:

- http://copyright.cornell.edu/resources/publicdomain.cfm

YouTube Info:

- http://www.thesitewizard.com/general/embed-youtube-video-copyright-matters.shtml
- http://www.youtube.com/t/copyright_education

Connections

Connect! Activity

It's important that all the images, graphics, and text on your blog are legal for you to use. If you've had your blog for any length of time go back and make certain you're in good shape legally.

Here are some things to check for:

- **Look for any old posts with song lyrics.** If the song isn't in public domain, take the lyrics down and replace them with a link to a YouTube performance of the song.

- **Check your images, especially any photos you've used.** Look for watermarks on them, which indicate they were used illegally.

- **Make sure you haven't produced any videos with songs that you don't own the copyright to.**

Steps to Protect Yourself from Hackers

Unfortunately, nothing is foolproof. Inevitably, the wiser we become at protecting ourselves, the more cunning those wishing us harm become. However, here are some tips to help prevent your accounts from being hacked.

The majority of times we get hacked are because we clicked a link that uploaded a virus which opened us up to hackers. **(Keep reading for information on what to do if your accounts have been hacked.)**

This is the bad news, but there's also good news. This kind of hacking is preventable, and here are some steps to take to stay safe online:

• **Be wise.** This seems basic, but so many times we just ignore our better judgment. How many of us have been sucked in by direct messages like these? "Have you heard the rumors your friend is spreading about you?" or "This is a hilarious video just uploaded about you." Stop. Think. Then don't click that link!

• **Assume it's a lie.** About six months ago I got an e-mail from an online company confirming a large purchase with my credit card. I knew I hadn't made any purchases, but still had to fight the urge to panic. I took a step back and looked more closely at the e-mail. I noticed several things that made me suspicious. I immediately did an online search for scams involving that company and came up with pages of recent victims. I contacted the company directly (not through the info in the e-mail) and confirmed the e-mail was a ruse.

Connections

- **Never give out sensitive information.** Let me repeat, NEVER GIVE OUT SENSITIVE INFORMATION! Companies don't ask for bank account info, passwords, or other information over the Internet. First, if you're a customer, they already have all of the information they need from you. Keeping up with personal passwords is a liability for companies.

- **Now, a quick word about passwords.** I know you don't want to hear it, but your password should be different for every account you have. If you're like me you probably have dozens of accounts, so how can you keep up with all those passwords? Trust me, it's not with sticky notes or a file on your computer.

Instead, take advantage of some wonderful programs. Some charge a small fee, others are free—all have the highest security rating available. **These are some options:**

- LastPass http://lastpass.com/index.php

- PassPack http://www.passpack.com/en/home/

- 1Password https://agilebits.com/onepassword

- KeepassX http://www.keepassx.org/ and Keepass http://keepass.info/

There are also blank booklets available for those of you who are hard-core old school and want something you can hold. I've seen them at local discount stores, as well as high-end specialty stores.

Connect! Activity

Now is a good time to get all your passwords stored in one secure place. As you're collecting them, change any that are the same.

HELP, I'VE BEEN HACKED!

Many of us have experienced the sinking feeling that comes with the realization that one (or more) of our social media profiles have been hacked. The easiest way to get hacked is to click on a link that contains a virus allowing hackers access to your accounts. When you click on the link, you may see a message telling you the link is broken. That doesn't mean you've dodged the bullet. If your firewall or virus software doesn't catch it, you'll still be infected.

One of the most common messages that spread viruses and allow hackers access is: "Did you know your friends are spreading nasty rumors about you?"

Or: "This is a hilarious video of you."

So what's a social media professional to do, cancel all your accounts and slink away in shame?

Never.

You need to repair the damage and keep it from happening again. To do this, first change your password for the social media profile that's been compromised.

But changing your password is just the first step. Now you need to check and make certain the damage hasn't spread. For this, you'll need to visit the app section for Twitter and Facebook. Here, too, you'll see a list of apps that have access to your Facebook account. Again, I recommend you revoke access to any you don't recognize or haven't used in a while. Doing these things should minimize the fallout that comes from being hacked.

Connections

Connect! Activity

Here is a short list to help you keep from getting hacked:

- On a PC, keep your virus software up to date.

- On any computer, keep all your programs up to date.

- Don't click on links in an e-mail unless you know for sure they're safe.

- Don't click on links in social media that claim to show you in a positive or negative light.

- Don't use the same password for different sites.

- Change your passwords every three to six months.

- Don't ever give out your password for a site, even if the request appears to come from that website. It's a liability for websites to keep your password on file, so they don't.

- If you're suspicious of a link, verify it before clicking. Either e-mail the sender, if it's personal, or go directly to the website and find the information there.

HOOTSUITE

Social networking is a lot easier when you use a few simple ancillary programs to help you get organized. One of my favorites is Hootsuite. Hootsuite is a scheduling program that allows you to set the dates and times when your social media updates appear in various networks. There are other programs that do this as well, like Buffer. The one I like best though is Hootsuite.

I personally use the free version of Hootsuite, although it does have some paid options. **But here's what I get in the free account:**

- 5 social profiles
- Basic analytics.
- 2 RSS feeds.

Here's a short list of what I like about it, and what it does for me:

- It's easy to schedule TWEETS and Facebook posts.
- You can copy and paste TWEETS into the update window with ease.
- It has two options for URL shortening.
- It keeps track of your @contacts to make tagging easier.
- It has a great get-acquainted video to help even the rawest beginner get started.

Connections

Every morning I spend about thirty minutes scheduling my main TWEETS and Facebook posts for the day. I usually line up at least 15-20 updates for the day. I try to schedule them evenly throughout the day so I can reach people in different time zones. **Here are some of the things I try to include daily:.**

•   I search my inbox for interesting articles from my favorite social media sites.

•   I post updates from my regular writing related blogs, including My Book Therapy, the Blue Ridge Conference site, Novel Rocket, and Chip MacGregor's blog. There are many others, but these are some of my favorites.

•   I look to see which of my friends' blogs have sent me an e-mail update and I try to highlight those.

•   Most days I include something about my blog or books, but lately I haven't TWEETed as much about myself. I haven't really needed to. A lot of the people I interact with on Twitter are as diligent about posting info about others as I am.

During the day, while I'm working, I check Hootsuite periodically to see who's reTWEETed or mentioned me.

I do try to do an @reply to anyone who highlights something I've done. This accomplishes two things—it gives them a mention, and it's a public way to say thank you.

A lot of people ask me if my schedule has been beneficial. You better believe it! I've reached the magic 2000 follower mark on Twitter and I've been adding between 15 and 20 followers a day. Beyond that, I don't spend much more than the 30 minutes every morning to schedule my TWEETS and Facebook posts.

Connect! Activity

If you don't already have a Hootsuite account, now is the time to set one up. Spend some time with it, clicking on the various icons and playing around with the settings. It's important you become familiar with this valuable social media resource.

CHERISH YOUR E-MAIL SIGNATURE LINE

Have you considered your signature line? If you're not certain what that is, it's the line or lines including or directly below your signature that is automatically generated within your e-mail program. **This one simple addition to my life has made it much easier for people to:**

- Find my blog.
- Buy my book.
- Follow me on Twitter and Facebook.

Here are the basics you should include.

- Your name and/or the name you write under.
- A link to your website.
- Your blog address.
- Your Twitter account name.
- Your Facebook Fan Page link.
- A link to your latest book (if you write books).

Warnings:

- Keep your signature line to a maximum of six lines. Anything more and people lose interest.
- Try not to add graphics, these take longer to load and the person who receives the e-mail has to click on a permission tab to see them, and most of us don't bother.

Connections

Connect! Activity

Check all your various e-mail outlets and make certain you have a signature designated for each one. **Here's what to check:**

- **Your e-mail homepage**—Whether you use Gmail, Yahoo, AOL, etc. Each has its own way to designate a signature.

- **Your smart phone**—You can add a signature to e-mail sent from your smart phone. You can even add a signature to your texts.

- **Your tablet**—If you own an iPad or any kind of tablet you'll need to set up a signature there as well.

- **Your e-book reader**—If you can send and receive e-mail on your e-book reader, you'll also need to check and make certain you have an e-mail signature designated there.

RESPECT THE E-MAIL

One part of writing for the Internet that a lot of people miss is the lowly e-mail. This method of correspondence has too long been neglected and ignored. Because it's possible to reach so many through this simple delivery system, we've somehow relegated it to the unimportant.

It is actually one of the most vital ways of reaching out to others.

I've listed just a few of the reasons for this neglect here:

The multitude of e-mail we send and receive each day has... well... made us lazy. We no longer pay attention to the common principles of grammar. We dash off missives full of sentence fragments, misspellings, and punctuation gone wild.

The conversational tone has reduced our correspondence to an undecipherable mélange of clichés and half-conveyed information. We run thoughts and sentences together with the assumption the reader will get enough of what we're trying to say, to do what we want them to.

The audience we commonly target with e-mail generally consists of close friends and associates, and we impose on their relationships with us to cover our haste. What we've forgotten is they're just as busy and the time required to decipher our messages is, at least an irritation, and at most an imposition.

Connections

So what's a busy person to do? Below is a list of common e-mail mistakes that only take a few moments to catch with a quick scan and can greatly increase the respect your correspondence, and you, receive.

Common E-mail Mishaps:

Misspelled names. This is basic and it amounts to a first impression for the e-mail, as well as for us, if this is the first time we've corresponded with the recipient.

The overuse of exclamation points. Okay, I admit it. It's an easy way to get our point across, and I'm definitely guilty of this. But really, the more often it's used the less effective it becomes.

Extra or inconsistent spacing between sentences. In this day and age (and for those of you who missed it—the past 15 years) sentences should only have ONE space between them, not two. I know this seems picky, but when you're glancing at a paragraph the inconsistency is quite noticeable.

The confusion of when to use "that" versus "who". The word "that" is used following a thing (or a group). While the word "who" is used when referring to a person.

Trying to say too much at once. I know we're busy, but slow down and use separate sentences (or even paragraphs) for addressing unrelated items.

Consistent information. This is a big one. When you're giving a list of information, like upcoming events, give the same information—in the same order—for each event. Otherwise your recipient is left scrambling for times and locations. It's also easier for us to accidently omit the needed information if it's listed haphazardly.

Incorrect use of hyphens, en dashes, and em dashes. Here's a quick grammar reminder:

- **Hyphens** (also incorrectly referred to as dashes) are used in compound words, or words that are linked together as one thing. Example: five-year-old girl.
- **En dashes are slightly longer and are used to separate numbers.** Example: from October 7–8. You format this by typing the number, space, dash, space, number, space.

- **Em dashes are the longest of the three and are used when you're expanding or modifying a statement.** Example: give the same information—in the same order—for each event. In Microsoft Word, you format this by typing the word, then dash, dash, next word and space. In Pages, you format this by pressing the shift, option, dash keys at the same time. Notice there is no space on either side of an em dash.

Incorrect use of ellipsis (...). This punctuation mark is used to denote a break in thought or speech. It is not interchangeable with the em dash.

Use of periods at the end of a bulleted list. According the AP Style Guide (the general guide that governs writing for the Internet) a bulleted list should include the use of a period at the end of each bullet point. There are levels of acceptance for this guideline, and most people don't add a period at the end of a one word bullet point. But, the important thing here is consistency. If you end even one of the bullet points with a punctuation mark—like a question mark—you must end them all with a punctuation mark.

Use of contractions. While this isn't a requirement, it is something that should be considered, especially if you're writing in a conversational tone. We speak in contractions, but we don't naturally write in contractions. After you've written an e-mail, give it a quick read—out loud—to find the places you would normally use a contraction.

Extraneous "that." This one simple word, so loved by high school English teachers, is horribly overused. Whenever possible, leave it out. For example:

- **Incorrect:** The teacup that I love best is an antique.
- **Correct:** The teacup I love best is an antique.

Common comma usage. A book could be written about how to correctly use commas, and most of it would be considered wrong by the next grammarian you consult. But there are still a few hard and fast rules. Here are two:

- When you have a compound sentence, connected with a conjunction, you use a comma if the second part of the sentence is a complete sentence.
- A serial comma is not used with the AP Style Guide.

Correct Example: red, white and blue.

Connections

Disorganized method of delivery. This one is more of a general concern. But it's important to construct your e-mail in an orderly fashion. Group topics in paragraphs and keep the flow of the information logical.

When we pay attention to these small issues, we convey an attitude of respect toward our audience. With just a small amount of effort, we show how we value their time, and it invites a respectful consideration of the information we're sending.

Connect! Activity

Pull up a couple of your old e-mails and take a minute to scan them, looking for these types of errors. Make your own list of common faux pas and keep it near your computer to help you remember what to look for.

One of the best parts of Twitter is the HASHTAG. This lowly number sign turns any bit of text into a searchable topic within the Twitter universe. But, the magic only happens if you choose your HASHTAGS wisely.

Today I'm going to review some of the best practices for using, finding and researching HASHTAGS.

HASHTAG Etiquette

Try to never use more than three HASHTAGS in any one TWEET. If you can make it two, that's even better. Otherwise you end up looking like a used car salesman. If you're trying to reach more groups, schedule multiple TWEETS, at different times, about the same subject and target your groups two at a time.

Always research your HASHTAG before you use it. Never assume it's the correct one. For example, I was targeting military families with TWEETS about my devotional for military families and I thought #military would be the logical HASHTAG. No, turns out that HASHTAG is frequently used by those trying to date someone in the military. Not really the demographic I was trying to reach. The HASHTAG I wanted was #militaryfamily and #deployment. Here are the four places I like to research HASHTAGS:

- http://Google.com
- http://HASHTAGS.org
- http://trendsmap.com
- http://whatthetrend.com

It's as Easy as A B C

Social Media doesn't have to be difficult. These tips can take you farther, faster, if you implement them into your social networking strategy.

A is for Answer. Think about the questions your target audience has, and use social media to give them the answers. For instance, my followers want to learn how to use social networking efficiently, so I TWEET shortcuts, tips, and easy solutions that take up very little time.

B is for Basics. Whatever you're doing, make it excellent. That means take time to check for typos and readability in your TWEETS, Facebook posts, and in your blog. Also be sure you're giving accurate links. There's nothing more frustrating than clicking on a link that takes you nowhere.

C is for Conversation. Remember social networking is all about interaction between people. It's NOT talking at them... it's talking with them.

D is for Direction. Think about what you're trying to accomplish and be deliberate.

E is for Effort. Social Networking feels like it should be easy. But learning a new language is never done without expending effort.

F is for Friends and Followers. These are the basis of your social interaction online. Don't neglect them and only chase new connections. Be true to this foundation and it will sustain you over the long haul.

Connections

G is for Google. Google alerts are one of your best social networking friends. You can set up alerts to let you know when someone has mentioned you or a subject you're interested in. And best of all, it's a free service. Just click Google Alerts (http://www.google.com/alerts) and follow the instructions.

H is for HASHTAGS. These are the gems in the fields of Twitter. Learn to follow HASHTAGS and use them correctly to increase your cyber-reach.

I is for Include. Include others more often than yourself when you send out TWEETS, posts and updates. The litany of me, me, me gets old fast in social networking.

J is for Journey. Social networking is all about the journey and the joy of discovery. It's NOT about arriving at a static destination.

K is for KISS. KISS is an old-fashioned acronym for "keep it simple silly." When you try to use too many different social networking tools you can get overwhelmed. Stick to three or four that work, and leave the dabbling to others who aren't working to become writers.

L is for Links. Used correctly, links or (hyperlinks) can increase your visibility and make your content much more valuable to those who follow you.

M is for Mobile. In the social networking world, more and more people view your blogs, TWEETS, and Facebook through a mobile device (phone, tablet, even e-reader). Make certain you're mobile friendly and all your sites are optimized for mobile viewing.

N is for Nice. Nice is the basic etiquette for social networking. My grandmother would have loved social media because the rules are simple—treat others the way you want to be treated.

O is for Original. Remember that original is who you are. There isn't another person like you anywhere. That is the essence of your value. The way you see things, share things, and interact is filtered through you. Don't be a copycat, be yourself.

P is for Permission. Just because you see it online doesn't mean it's yours to borrow. Remember to respect the rights of others and take time to learn about copyrights.

Q is for Quit. That's right, quit. Set a time limit for social networking and then quit when it's reached. You're trying to be a writer, not a full-time marketing manager.

R is for Reach. Always look for new places to reach new people. Comment on a new blog or follow a friend of a friend.

S is for Start. Many people put off social networking because they think it's hard or unpleasant. You'll never know till you try (sorry, that's my grandmother again).

T is for Tease. When you compose a headline, TWEET or Facebook post, don't sum up your message. Tease your audience with something compelling that will encourage them to look further or interact with you and each other. In other words, become a social networking flirt!

U is for Understand. Take time to understand the social networking world. Learn the etiquette involved before you run headlong into embarrassment.

V is for Value. This is ALWAYS the basis of your message, no matter your social networking medium. Bring value to your followers, and they'll be friends for life.

W is for Wait. There are lots of people out there who are trying to sell you a shortcut to friends and followers. The problem is that audiences are fickle. Take time to build your network through value and personal relationships, and your numbers will mean something—to you and to potential publishers.

X is for X-ray. Make certain the bones of your social networking strategy are strong. Take time to build a solid framework, and you'll spend less time and get more results.

Y is for YouTube. Videos are popular on Twitter, Facebook, and in Blogs. Learn to leverage the use of videos to add value and fun to your relationships.

Z is for Zoo. Yes, the whole social media thing is a bit of a zoo. I tell people truthfully, you're doing something right if you always feel like you're behind in learning about social media. It doesn't matter who you are, NO ONE can know everything about social media—the field is just expanding too quickly.

GETTING IT ALL DONE: A TESTIMONY

You can look at this digital age we live in as a blessing or a curse—and there are merits to each viewpoint. Since I'm a glass-half-full type girl, I happen to take the positive approach. I like the connectedness of this time and place. I like connecting in person and online—especially through Facebook and Twitter.

A word of warning here, it's possible to let these tools (Facebook, Twitter, Pinterest, etc.) eat into your writing time. For myself, the way I combat that temptation is to schedule my networking time.

Many of you have asked how I schedule my day so now is as good a time as any to share it with you. Now, please know that I am NOT an organized person, but this loose arrangement of my day helps me to stay sane in the insane world of writing.

•   8 – 9: I answer e-mail (I have two accounts), and I use Hootsuite to schedule my main social media for the entire day. I use this hour to get connected.

•   9 – 11:30: I use these hours for creative writing because it's the time when I'm most inspired.

•   11:30 – 12: I answer any e-mails and phone calls that have come in and again check FB, Twitter and my blog

•   12-1: lunch

•   1-3: I work on things that have a deadline, and once a week I write all my blog posts during this time.

•   3:30 – 4: I again check e-mail, phone messages, FB, Twitter, and my blog.

Connections

Then, before I go to bed I again check e-mail, FB, Twitter, and my blog.

Also, about once an hour I get up and walk around to relieve my back and when I sit back down, I check Hootsuite. That way, if anyone has mentioned me or reTWEETed something, I can reply. It's important to keep the conversation going throughout the day, and this is how I do it. BUT, I only allow myself 5-10 minutes each hour or two.

Now, some days the times vary, especially if I have a big deadline, but this is my basic schedule. I try not to be subject to e-mails or even the telephone. Funny thing, I discovered that people are fine about having to leave messages, IF they know I'll actually call them back. This allows me to accomplish what I need most days and still stay sane.

DON'T GET BLOGGED DOWN

This may seem like an odd topic to include in this book, but I've gotten a lot of questions lately about how some folks are following so many blogs they don't have time to write. Of course I'm also hearing from some who haven't committed to following any blogs because of a fear that there won't be any time to write.

Either extreme will blog... er... bog down your career.

All kidding aside, as a writer it's vitally important to continue your education. And blogs can be an extremely efficient way to do this. I can hear the chorus of complaints now:

There are millions of blogs out there—how do I find one that's worthwhile?

How do I narrow down my choices—do I have to read hundreds to find one that fits what I need?

Where do I start looking—it's gonna be like finding a needle in a haystack!

Enough already.

Finding high quality blogs that pertain to your situation are just not that hard to find. You find a good blog the same way you find a good book. Think about it: we've been shopping for books for years, choosing from millions of titles, and finding success.

Connections

Here are some steps for finding blogs that you will want to follow:

- First, look for recommendations from people you trust.

- Second, most blogs have blog lists as a part of their layout. If you like blogger A and he likes blogger D, chances are you ought to give blogger D a try.

- Third, look for blogs from familiar people. Nowadays, most bloggers have or are at least affiliated with more than one blog.

So how do you manage following blogs and keep from eating into valuable writing time?

I like to use a three-tiered approach.

- I start with three to five blogs that I follow regularly. These are blogs I almost always read.

- Then I have a second level of six to eight blogs that I watch closely. I read these about one half of the time and I decide when by paying close attention to the subject lines and titles of the posts.

- Finally I have a third level of about 12–15 blogs that I watch the subject lines and titles and read when they sound interesting.

WHY SOCIAL MEDIA ISN'T SO HARD AFTER ALL!

Social media is a return to a simpler age.

I can see the skepticism on your faces from here, but bear with me. I think you'll see the connection.

First, I'd like to invite you to remind yourselves of the standards I, and most of you, were taught growing up. We were raised by certain ideas about how to treat others. My mother and grandmother had a name for it—polite society. **Here are some of the basics, in case you've forgotten**:

- If someone says something nice about you, thank him.
- When someone does something nice for you, do something nice for her.
- Always put others before yourself.
- If you can't say something nice, don't say anything at all.

It was a "Do unto others as you would have them do unto you" world.

These rules guided my behavior in almost all circumstances. And they made the world I lived in pretty easy to navigate. We all operated from a common basis, and everyone knew what was expected from everyone else.

These same basic rules are once again enjoying resurgence—**on the Internet.**

Connections

Stay with me and consider our interaction on social media, specifically Facebook and Twitter.

- If someone mentions you (which is a nice thing in this new, platform-building paradigm) you thank him.

- When others do something nice for you online, like telling people you have a great blog post, you tell your friends about their blogs.

- To keep from becoming a self-centered sounding boor, promote others online more than yourself. I know it's counterintuitive, but it works every time. Those who promote others are always more popular and have more friends than those who are self-serving and self-promoting.

- And most important of all, when almost everything ever said online can still be found somewhere online—NEVER share an update that puts someone else down.

It's once again a golden-rule governed world.

When I realized the relationship between how I was raised and this new frontier, I also saw that I have a lot of experience I can share with the younger, sometimes more digitally-familiar generation. And this gave me the confidence to embrace this new culture. Because let's face it, there truly is nothing new under the sun.

See, it's not so hard after all!

Thank you for reading,

Edie Melson

27397917R00085

Made in the USA
Columbia, SC
23 September 2018